Nuclear Physics 2

*I dedicate this book to my
darling baby Mariama*

Nuclear Physics 2

Radiochronometers and Radiopharmaceuticals

Ibrahima Sakho

WILEY

First published 2024 in Great Britain and the United States by ISTE Ltd and John Wiley & Sons, Inc.

ISTE Ltd
27-37 St George's Road
London SW19 4EU
UK

www.iste.co.uk

John Wiley & Sons, Inc.
111 River Street
Hoboken, NJ 07030
USA

www.wiley.com

Any opinions, findings, and conclusions or recommendations expressed in this material are those of the author(s), contributor(s) or editor(s) and do not necessarily reflect the views of ISTE Group.

Library of Congress Control Number: 2024931599

British Library Cataloguing-in-Publication Data
A CIP record for this book is available from the British Library
ISBN 978-1-78630-733-0

Contents

Preface

Nuclear physics is the study of the properties of atomic nuclei. Its aim is to understand the properties of *nucleons* and the mechanisms of *nuclear reactions* (spontaneous and induced), with a view to describing the various processes of *elastic* and *inelastic* nucleus–nucleus interactions.

Radionuclides are useful in many areas of everyday life: archaeology, biology, agronomy, medicine, industry, etc. Among the most spectacular are applications in radiochronometry (dating of archaeological objects, sediments and soils for the detection of anthropogenic pollutants, etc.) and nuclear medicine (radiopharmaceuticals used in nuclear medicine imaging, radiotherapy, etc.). The production of electrical energy in nuclear power plants exploits the properties of nuclear fission reactions. In addition, the study of nuclear physics enables us to understand many astrophysical phenomena, such as nucleosynthesis processes (primordial, stellar, explosive) within the framework of the Big Bang model. The study of these processes allows understanding of the origin of chemical elements and how to model the evolution of stars from their birth to their explosive end, for example, in supernovae and neutron stars [SAK 22].

This book, entitled *Nuclear Physics 2: Radiochronometers and Radiopharmaceuticals*, is divided into four chapters.

Chapter 1 is devoted to a description of the Big Bang model, enabling us to understand the origin of all known chemical elements from nucleosynthesis reactions. The chapter begins with a presentation of Christian Doppler's (1804–1853) theory, which led to his hypothesis of the physical phenomenon known as the *Doppler effect* in the case of *sound waves*. This is followed by a description of Christoph Buys-Ballot's (1817–1890) historic experiment confirming the Doppler effect. The study then turns to Armand Hippolyte Fizeau's (1819–1896) theory of the Doppler effect as it applies to light waves. This study establishes the formula for the Doppler–Fizeau effect, based on the transformation laws of the *wave quadrivector*. This is followed by a study of the longitudinal and transverse Doppler effects in the classical approximation for weakly relativistic motions, and the derivation of the Doppler–Fizeau formula used to interpret the *redshift phenomenon* of light sources in relative motion, and to calculate the radial velocity of a star or galaxy. In addition, the link between the sign of the Doppler shift and the relative motion of a light source is studied in the classical approximation, to highlight the phenomenon of redshift in the spectrum of stars and galaxies in particular. Following these developments, the principle of redshift measurement is illustrated schematically using the spectrum of the galaxy named NGC 3627 (NGC: New General Catalogue). Following this study of the redshift interpreted by the Doppler–Fizeau effect, the chapter then turns to the theoretical and experimental facts that have validated the Big Bang model. These range from the observation of the redshift to the discovery of the *cosmic microwave background* (CMB). This historical overview begins with the first observations of the redshift phenomenon in the spectral lines of galaxies by de Vesto Melvin Slipher (1875–1969). This is followed by the work of Alexander Friedmann (1888–1925), who first published a theory of the expansion of the universe. The work of Georges Lemaître (1894–1966), linking the expansion of the universe with observations of the escape velocity of extragalactic nebulae, and his formulation of the "primitive atom" hypothesis to explain the origin of the universe by introducing the notion of instant zero, figure prominently in this historical review. The presentation of this work is followed by Edwin Hubble's (1889–1953) decisive observations showing that the variation of velocity with distance is linear, a relationship known as the Lemaitre–Hubble law. Various attempts to estimate the Hubble constant denoted H_0 are then discussed. The discovery of the cosmic microwave background, the decisive argument in favor of the Big Bang theory, ends this historical review. This is followed by a brief description of the chronology of the universe's evolution after the Big Bang. The various eras that characterize the chronology of the universe are studied: the Planck era, the era of grand unification, the era of inflation, the era of baryogenesis and primordial nucleosynthesis, the era of quark–gluon plasma formation, the era of nucleosynthesis, the dark age of the universe, designating the era beginning with *radiation–matter decoupling*, the radiative era and finally the era of star and galaxy formation. The chapter is interspersed with corrected application exercises.

Chapter 2 is reserved for the study of the *various nucleosynthesis processes* that began almost 1 second after the Big Bang and lasted approximately 3 minutes. The chapter begins with an overview of the concept of chemical elements, of which there are 118 known, 90 of which occur naturally on Earth. This is followed by a detailed study of the processes of primordial, stellar and explosive nucleosynthesis. The study of primordial nucleosynthesis describes the formation of light elements such as hydrogen, deuterium, helium-3, helium-4, lithium-6 and lithium-7 in the first instants of the universe. The study of stellar nucleosynthesis enables us to understand the origin of carbon-12, oxygen-16, neon-20, sodium-23, magnesium-24, silicon-28 and 30, sulfur-31 and phosphorus-30 and 31, as well as that of all nuclei up to iron-56. We then study the formation of all elements heavier than iron and isotopes synthesized via the s (slow) and r (rapid) processes during explosive nucleosynthesis. This study enables us to describe the "slow" neutron capture process via the s process, as well as the rapid process of radiative neutron capture followed by decay, which provides about half the abundance of elements beyond iron up to uranium. This chapter also covers the spallation process, corresponding to the formation or destruction of large nuclei by very high-energy particles (such as the nucleosynthesis of Li, Be and B in the interstellar medium), and the photodisintegration process, which reflects the destruction of nuclei by photons. The study then focuses on the description of important nucleus-forming processes, such as the triple-alpha reaction, a set of nuclear fusion reactions simultaneously transforming three α particles (helium-4 nuclei) into carbon-12 nuclei via the unstable beryllium-8 nucleus. We also look at the formation of compound nuclei, in particular, the ^{14}N (p, $\gamma)^{15}O$ reaction involved in the *CNO* (Carbon–Nitrogen–Oxygen) or the Bethe–Weizsäcker cycle studied in astrophysics. Finally, the chapter focuses on the classification of natural and artificial radionuclides in the environment. The chapter is also interspersed with corrected application exercises.

Chapter 3 is dedicated to the study of *radiochronometers applied to dating*. It begins with a study of the principle of carbon-14 dating. This introduces the notions of cosmogenic isotopes, cosmic radiation and calendar age. It introduces the notions of the "Bomb" effect and the "Suess" effect, which contribute to modifying the concentration of radiocarbon in the atmosphere. It also introduces the notion of the reservoir effect, reflecting the fact that oceanic and atmospheric concentrations of radioactive ^{14}C are not homogeneous. Next, the study focuses on the principle of potassium–argon (K–Ar) dating. This method establishes the age equation of a volcanic eruption, only taking into account the ^{40}Ar resulting from the decay of the ^{40}K present in the lava (this argon 40 is often referred to as $^{40}Ar^*$). Next, the age equation is corrected to take account of the ^{40}Ar atmosphere, so that the results obtained with the K–Ar clock can be properly used. This is followed by a description of the principle of dating soils or sediments using the radiochronometers lead-210, cesium-137 and beryllium-7. A description of the principle of lead-210 dating explains the origins of supported lead-210 ($^{210}Pb_{sup}$) and excess lead-210 ($^{210}Pb_{ex}$) in sediments. Next, the CFCS (Constant Flux and Constant Sedimentation), CRS (Constant Rate of Supply) and CIC (Constant Initial Concentration) models are described, enabling the age of a sediment to be determined experimentally. This chapter features a study of the atmospheric nuclear tests carried out between 1945 and 1980, and of the Chernobyl accident in 1986, the second largest source of cesium-137 in the atmosphere. This is followed by a description of the principle of ^{137}Cs radiochronometer dating. This involves taking core samples from the sediment in question and interpreting the ^{137}Cs activity profile, according to the sampling date.

This is followed by a description of the principle of dating using the cosmonuclide ^{7}Be, formed in the troposphere by nuclear spallation. The chapter concludes with a description of the principle of dating using the uranium–thorium radiochronometer to determine the age of certain carbonate formations of animal or sedimentary origin, and the principle of dating using the uranium–thorium and uranium–protactinium radiochronometers for coral dating. As in previous chapters, corrected application exercises are provided at various points in the chapter.

Chapter 4 is devoted to *general information on radiopharmaceuticals used in nuclear medicine imaging*. The chapter begins with a definition of nuclear medicine and the aims of this discipline. This is followed by a description of the various fields of application of nuclear medicine, and then a brief overview of the birth of nuclear medicine, from the *first* use of radioisotopes as tracers in plant biology in 1913 to the development of positron emission tomography (PET) in 1975. Following this genesis, the different types of diseases diagnosed in nuclear medicine are examined. These include cardiovascular diseases, cancers and neurodegenerative disorders, such as Alzheimer's, Parkinson's and Lewy body dementia. This is followed by a general introduction to cancer, focusing on cellular organization in the body and the evolution of cancer cells, leading to the notion of tumor and the formation of metastases. This development introduces the notions of carcinogenesis (or oncogenesis). Next, a description of normal and tumor angiogenesis is given, introducing the vascular endothelial growth factor (VEGF). This biological factor plays an essential role in normal and pathological vasculogenesis and angiogenesis. Following this description, the development focuses on global cancer epidemiology data between 2018 and 2023, as well as recommendations from cancer agencies. Then, the specific properties of radiopharmaceuticals are studied, in particular, the process of synthesizing radiopharmaceutical drugs. The quality control of radiopharmaceuticals is examined, introducing the concepts of radiochemical purity, radionuclidic purity and abnormal toxicity testing. This is followed by a description of the various experimental methods for determining radiochemical purity, such as thin layer chromatography (TLC), column chromatography (CSC) and high-performance liquid chromatography (HPLC). Following this development, the principles of PET and single-photon emission tomography (SPECT) are described. The chapter then goes on to describe the different radioisotopes used in nuclear medicine imaging, as well as the PET-scan or TEP-scan (Positron Emission Tomography, coupled with a scanner). Finally, the chapter closes with a presentation of the main scintigraphies and their uses in nuclear medicine.

The four chapters are followed by three appendices devoted to a study of neurodegenerative dementia, in relation to the content of section 4.6 of Chapter 4. These examine the properties of the radiopharmaceutical ^{123}I-ioflupane, used for differential diagnosis between essential tremor and neurodegenerative diseases, such as Parkinson's disease, Lewy body dementia and Alzheimer's disease. Detailed explanations are given of the causes and effects, risk factors and diagnosis of Alzheimer's disease (Appendix 1), which is responsible for cognitive and behavioral disorders in 35 million sufferers according to studies carried out in 2022, Lewy body dementia (Appendix 2), which is very common and accounts for approximately 20% of dementia cases, and Parkinson's disease (Appendix 3), with a prevalence of over 2% after the age of 65, according to studies carried out in 2020. These appendices are followed by an extensive bibliography, enabling readers to deepen their knowledge of the book, which is rounded off by an index.

We would like to express our gratitude to Prof. Maurice NDEYE, Director of Research and Head of the Carbon-14 Laboratory at the Institut Fondamental d'Afrique Noire Institut Fondamental d'Afrique Noire, Cheikh Anta Diop University, Dakar, Senegal, for his valuable review notes on radiocarbon 14 dating. Likewise, our warm thanks to Dr Frédéric Thévenin, Astrophysicist at the Observatoire de la Côte d'Azur – Nice, France, *for his invaluable corrections and remarks on nucleosynthesis processes.*

This book is written for pupils, physical science teachers, students, teacher-researchers and professionals working in the fields of astrophysics, dating-related environmental sciences and nuclear medicine.

This book is written in clear, concise language, with a typographical structure similar to that of Volume 1. Each chapter begins with a presentation of the general objective, the specific objectives and the prerequisites for understanding the chapter to be studied. In addition, each chapter is interspersed with simple application exercises for a good understanding of the properties of the radioelements studied.

This book does not attempt to cover every aspect of radioelement origins and applications in radiochronometry, nor of the duality between nuclear medicine and radiopharmaceuticals. As human work can be improved, we are always ready to listen to our readers' suggestions, comments and criticisms, which could help improve the scientific quality of this book.

February 2024

A Description of the Big Bang Model

Overall objective	
Understand the chronology of the universe's evolution from the Big Bang to the formation of stars and galaxies.	
Specific objectives	
Explain the Doppler effect for sound waves;	Know the relationship between the four fundamental interactions in the universe during the Planck era;
Interpret the Buys-Ballot experiment that confirmed the Doppler effect;	Retain the special case of the evolution of gravity during the Grand Unification era;
Explain the Doppler–Fizeau effect;	Justify the unification of interactions during the Grand Unification era;
Establish the formula for the Doppler–Fizeau effect in the relativistic case;	Consider the special case of the evolution of the strong nuclear interaction during the inflationary era;
Express the formula for the Doppler–Fizeau effect in the classical approximation;	Illustrate schematically the evolution of the universe from the singularity to the instant of total separation of the four interactions at the temperature of the cosmic microwave background;
Calculate the radial velocity of a star or a galaxy;	Define the baryogenesis era;
Illustrate schematically the phenomenon of red-shifting the spectrum of a light source;	Retain the asymmetry between matter and antimatter during baryogenesis;
Define the Doppler shift in the classical approximation;	Explain the chemical composition of the universe during the era of nucleosynthesis;
Explain the relationship between the sign of the Doppler shift and the relative motion of a light source;	Justify the dark age of the universe;
Explain the principle of measuring redshift using the spectrum of a given galaxy;	Make the connection between the dark age of the universe and the period of recombination;
Define the Cosmic Microwave Background (CMB);	Justify the radiative era of the universe;
Retain FDC temperature value (2.7 K);	Know the formation period of stars and galaxies;

Learn the main theoretical and experimental facts that led to the validation of the Big Bang model, from the observation of the redshift to the discovery of the cosmic microwave background;	Know the mechanism of star formation;
Understand the evolution of the universe from the singularity to the age of inflation;	Know the particularity of the main sequence in the Hertzsprung-Russell diagram.
Prerequisites	
Sound waves;	Constitution of atomic nuclei;
Spectrum of the hydrogen atom;	Isotopy;
Balmer's empirical law;	Nuclear reactions;
Lorentz spatial transformations;	Alpha, beta and gamma radiation.

1.1. Red-shift phenomenon in the spectrum of stars and galaxies

1.1.1. *Doppler effect*

In 1842, the Austrian physicist Christian Doppler (1804–1853) explained that our perception of the pitch of a sound is altered by the relative motion of the sound source to the observer. He then suggested that the colors of stars could be due to a similar effect, affecting their light [PAT 12]. He presented his theory to the Royal Bohemian Academy of Sciences in Prague. This phenomenon is known as the *Doppler effect* in the case of *sound waves*.

Consider a sound wave emitted by a moving source (*train whistle, ambulance siren, etc.*) relative to a stationary observer. The sound perceived by the observer's ear becomes increasingly *acute* (shift towards higher frequencies) as the source approaches. As the source moves further away, the perceived sound becomes increasingly *low-pitched* (shift towards shorter frequencies). *Sound frequency corresponds to the number of vibrations per second of a sound. The average human ear perceives sounds in a certain frequency range, from approximately 20 Hz to 20,000 Hz. Low frequencies* range from 20 Hz to 200 Hz, *mid or medium frequencies* from 201 Hz to 2,000 Hz, and *high frequencies* range from 2,001 Hz to 20,000 Hz. For a given sound, there are several frequencies at a distance, including the *fundamental frequency* f_0 and higher frequencies, called *harmonics* $2f_0$, $3f_0$, etc.

NOTE.–

Ultrasound is sound vibration at frequencies above 20,000 Hz. These are sound waves that cannot be detected by the human ear. However, they are audible to certain animals, such as bats, dolphins and whales.

For French speakers, the *pitch of a sound* is designated by the name of a note on a scale. Today, there are seven notes: do, re, mi, fa, sol, la, si, corresponding to predetermined frequencies. English speakers use the letters A, B, C, D, E, F, G respectively.

In addition, a note corresponds to several frequencies. To distinguish the *frequencies of a note*, they are numbered using numbers called octaves, with values varying from -1, 0, 1, 2, 3, 4, 5, 6, 7, 8 and 9. For example, to the note *la* correspond the frequencies 27.5 Hz, 55.0 Hz, 110 Hz, 220 Hz, 440 Hz, 880 Hz, 1,760 Hz, 3,520 Hz, 7,040 Hz and 14,080 Hz. These frequencies correspond respectively to the octaves from -1 to 8.

At the international conference in London in 1953, *la3 at 440 Hertz* (440 vibrations per second) *became the reference for all the world's orchestras*. Table 1.1 shows the frequencies of some musical notes.

Note	sol3	sol#3	la3	la#3	Si b3	si3	do4	do#4
Frequency (Hz)	392	415	440	466	466	494	523	554

Table 1.1. *Frequencies of some musical notes ("#" symbol for diese and "b" symbol for bemol). For a color version of this table, see www.iste.co.uk/sakho2/nuclearphysics2.zip*

The Doppler effect was first subjected to experimental verification in 1985 by the Dutch physicist Christoph Buys-Ballot (1817–1890), who had musicians play in a train moving at speed *v* (Figure 1.1).

Figure 1.1. *French schematic illustration of the Buys-Ballot experiment. Trumpeters play a "la" in a moving train. Other musicians standing still along the platform listen to the note played and clearly hear a "la #". Source: http://sosphysique.ac-poitiers.fr/ viewtopic.php?t=6833, 2016. For a color version of this figure, see www.iste.co.uk/ sakho2/nuclearphysics2.zip*

In his experimental study, Buys-Ballot placed musicians on a train and asked them to play a la3. Musicians, placed on the platform at regular intervals, were able to finely distinguish the differences in note pitch. When the train approached, the musicians on the platform claimed to have heard a "la#", i.e. a note a semitone higher (the smallest difference between two notes of a given scale).

Looking at the data in Table 1.1, we see that the note "la#" corresponding to the frequency 466 Hz is shifted from the reference frequency 440 Hz of the note la3. This shift $\Delta v = 26$ Hz towards the high frequencies was a clear confirmation of the Doppler effect. In the Buys-Ballot experiment, this shift was used to estimate the train's translational speed. The value of the frequency of the "la#" can be found by applying the *Doppler–Fizeau formula* (see Application 1.1).

APPLICATION 1.1.–

> June 3, 1845 [...]: the Maarsen line is decommissioned for scientific purposes. By order of the Minister of the Interior, three musicians set up in the open carriage, tuning their piston horns, and several groups of observers, equipped with an absolute ear, a notebook and a pencil, are scattered at specific intervals along the track. As soon as the train reaches its maximum speed, i.e. over 70 km h−1, the musicians will play, in unison and as loud as possible, so as to drown out the noise of the locomotive, a la which the observers placed along the track will try to estimate the alteration, in half and quarter tones, under the effect of speed [...]. The experiment was repeated two days later with a slightly different protocol (you never know), different speeds and different instruments (trumpets), all just as carefully tuned, and it produced the results everyone was expecting [...]. Riding his locomotive hard, Buys-Ballot demonstrated – with such panache! – what everyone had taken for granted [WIT 03].

Using the data from the previous section and the introductory text, show that the Buys-Ballot experiment confirms the Doppler effect. The formula for the Doppler–Fizeau effect applied to sound waves is given (see demonstration at the end of section 1.1.2):

$$f = f_0\left(1 - \frac{v}{c}\right)$$

[1.1]

The speed of sound: $c = 340$ m·s^{-1}.

ANSWER.–

Using [1.1], we derive the expression for the translational speed v of the train. Let:

$$v = \left(1 - \frac{f}{f_0}\right)c \qquad \qquad [1.2]$$

The trumpeters played *la3* at frequency $f = 440$ Hz. The musicians at rest on the platform heard the note *la#3* at frequency $f_0 = 466$ Hz (in the terrestrial reference frame at rest).

$$v = \left(1 - \frac{440}{466}\right) \times 340 = 18.97 \text{ m·s}^{-1} = 68.3 \text{ km·h}^{-1} \approx 70 \text{ km·h}^{-1}. \qquad [1.3]$$

The value 68.3 km h^{-1} is in very good accordance with the minimum velocity 70 km·h^{-1} indicated in the preliminary text. The Buys-Ballot experiment confirms the Doppler effect.

1.1.2. *Doppler–Fizeau effect*

In 1848, French physicist and astronomer Armand Hippolyte Fizeau (1819–1896) demonstrated that the velocities of stars are far too low compared to the speed of light to cause any appreciable change in their colors. He also concluded that we could expect to detect slight variations in the wavelengths of the lines in their spectrum. The phenomenon of the *redshift* of any light source (star, galaxy, etc.) is illustrated in Figure 1.2 [SAK 23a].

The velocity of a star relative to the Sun can be broken down into two components: a velocity on the celestial vault, called *tangential velocity* v_t, and a velocity on the star's line of sight, called *radial velocity* v_r.

The *proper motion* (μ) is the tangential angular displacement of a star on the celestial vault. The proper motion is obtained by comparing the positions of stars in photographs taken several years apart, and is expressed in arc seconds per year ("/year), with $1'' = 1/3{,}600°$. To obtain the tangential velocity of a star from its proper motion, we also need to know its distance [DUP 01]. The tangential velocity v_t, expressed in km h^{-1} is given by the formula: $v_t = 4{,}74\mu d$. The radial velocity is determined by the Doppler–Fizeau formula (see Application 1.2).

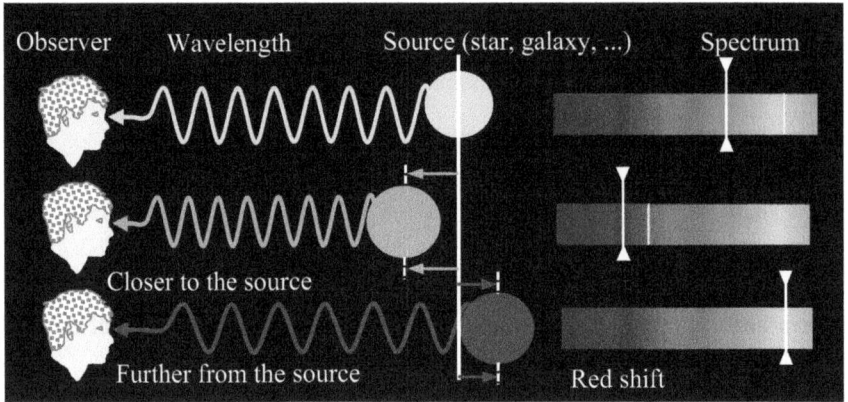

Figure 1.2. *Schematic illustration of the redshift in the spectrum of a light source in relative motion with respect to a fixed observer. For a color version of this figure, see www.iste.co.uk/sakho2/nuclearphysics2.zip*

NOTE.–

i) Ignorant of Fizeau's work, Austrian physicist and epistemologist Ernst Mach (1836–1916) carried out his first work on the *Doppler effect, optics* and *acoustics*. In 1860, he reached the same conclusions as Fizeau (1848), i.e. 12 years after. In the literature, however, it is the Doppler–Fizeau effect that is used.

ii) Seen from Earth, at a distance of 8.611 light-years, *Sirius*, located in the constellation of the Great Dog, is the brightest star in the sky after the Sun. It is a binary star (made up of two stars, Sirius A and Sirius B, orbiting about 3 billion kilometers apart).

Let us now establish the Doppler–Fizeau formula in the special relativity. We will deduce the classical expression valid for low speeds, relative to the speed of light.

To explain this effect, let us consider a quantum system S at rest in a reference frame R_0 in uniform rectilinear motion with velocity v relative to an observer O linked to a reference frame R (Figure 1.3).

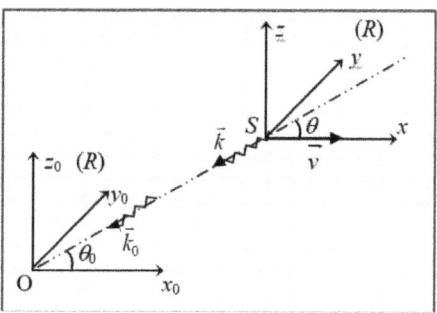

Figure 1.3. *A light source S moving at speed $\bar{v}$ emits a wave with the wave vector k in the reference frame R0. In the reference frame R linked to the observer O, the wave vector is k_0*

Using the Lorentz spatial transformation, we deduce the transformation law of the *wave quadrivector* coordinates ($\vec{k}$, ω/c); we get:

$$\begin{cases} k_x = \gamma_{(v)}\left(k_{0x} - \beta\omega_0/c\right) \\ k_y = k_{0y} \\ k_z = k_{0z} \\ \omega/c = \gamma_{(v)}\left(\omega_0/c - \beta k_{0x}\right) \end{cases} \qquad [1.4]$$

In the transformation relations [1.4], $\beta = v/c$ and $\gamma_{(v)} = (1 - \beta)^{-1/2}$.

Knowing that the projection k_{0x} of the wave vector along the Ox axis in the R_0 reference frame is equal to $k_{0x} = -k_0\cos\theta_0$, the fourth relation of the transformations [1.4] gives:

$$\frac{\omega}{c} = \gamma_{(v)}\left(\frac{\omega_0}{c} + \beta k_0 \cos\theta_0\right)$$

Replacing $\gamma_{(v)}$ and β by their expressions, we finally find ($k_0 = \omega_0/c$):

$$\omega = \omega_0 \frac{1 + \dfrac{v}{c}\cos\theta_0}{\sqrt{1 - v^2/c^2}} \qquad [1.5]$$

– for $\theta_0 = 0$, equation [1.5] leads to the formula for the *longitudinal Doppler effect*:

$$\omega = \omega_0 \frac{1 + \dfrac{v}{c}}{\sqrt{1 - v^2/c^2}}$$ [1.6]

– for $\theta_0 = \pi/2$, equation [1.5] leads to the *transverse Doppler effect*:

$$\omega = \omega_0 \frac{1}{\sqrt{1 - v^2/c^2}}$$ [1.7]

For weakly relativistic movements ($v \ll c$), the longitudinal and transverse Doppler effects can be written as follows:

– for $\theta_0 = 0$, equation [1.5] gives the first order in v/c:

$$\omega = \omega_0 \left(1 + \frac{v}{c}\right)$$ [1.8]

– for $\theta_0 = \pi/2$, equation [1.5] gives $\omega \approx \omega_0$: the transverse Doppler effect disappears in the classical approximation.

In terms of frequency, equation [1.8] is written:

$$f = f_0 \left(1 + \frac{v}{c}\right)$$ [1.9]

As the source moves closer to the observer, we need to change the "+" sign in the parenthesis of equation [1.10] to the "−" sign. This is equivalent to changing v to $-v$ (or to $k_{0x} = k_0 \cos\theta_0$, in the fourth relation of [1.4]). We then obtain:

$$f = f_0 \left(1 - \frac{v}{c}\right), v > 0$$

We thus find the relationship [1.1] in the case of the Buys-Ballot experiment, remembering that $\omega = 2\pi f$ and $\omega_0 = 2\pi f_0$.

1.1.3. *Doppler shift expression*

In the classical approximation ($v \ll c$), the quantity z given by the relation [COM 12] is called the *Doppler shift*:

$$z = \frac{\lambda - \lambda_0}{\lambda_0} = \frac{v}{c} .$$ [1.10]

– λ: measured wavelength (source in motion) corresponding to frequency $f = c/\lambda$;

– λ_0: measured wavelength (stationary source) corresponding to frequency $f_0 = c/\lambda_0$;

– v: speed of the source relative to the fixed observer;

– c: speed of light in a vacuum ($c = 300{,}000$ km·s^{-1}).

The Doppler shift [1.10] shows that:

– as the source moves away from the observer, the measured wavelength increases and the Doppler shift is positive ($\lambda > \lambda_0$ and $f < f_0$) $\Rightarrow v > 0$;

– as the source moves closer to the observer, the measured wavelength decreases ($\lambda < \lambda_0$ and $f > f_0$) and the Doppler shift is negative $\Rightarrow v < 0$.

Italian astronomer Angelo Secchi (1818–1878) and English astronomer William Huggins (1829–1910) were the first to attempt the visual measurement of the wavelength shift in a star's spectrum, predicted by the Doppler principle. In 1871, Huggins announced that he had been able to measure the velocity of Sirius. However, Huggins' measurements were fraught with uncertainty. For example, for the star Sirius, he published a radial velocity of -40 km/s; whereas, the value measured today is -7.6 km·s^{-1} (see: https://www.techno-science.net/Alpha-Canis-Majoris-page-2.html). The radial velocity of Sirius compared with the value -7.6 km·s^{-1} will be calculated in Application 1.2.

The Doppler effect is also used in medical imaging, notably in Doppler ultrasound. The latter is used to analyze fluids (particularly blood) and tissue irrigation. In this technique, the Doppler effect is used to quantify circulatory velocities, while ultrasound is used to visualize vascular structures (see Appendix 2).

APPLICATION 1.2.–

The hydrogen spectral line, with a wavelength of 656.279 nm in the laboratory, has a wavelength of 656.263 nm in the Sirius star spectrum [TRE 22]. Demonstrate the Doppler shift expression [1.10] using [1.9]. Determine the radial velocity v_r of Sirius. Interpret the sign of the result obtained.

Given: speed of light in a vacuum: $c = 300{,}000$ km·s^{-1}.

ANSWER.–

– Demonstration of the Doppler shift or spectral shift:

Using [1.9], we obtain knowing that $\omega = 2\pi c / \lambda (v \ll c)$:

$$\frac{1}{\lambda} = \frac{1}{\lambda_0}\left(1 + \frac{v}{c}\right) \Rightarrow \frac{\lambda}{\lambda_0} = \left(1 - \frac{v}{c}\right)^{-1} \approx 1 + \frac{v}{c}.$$

That is:

$$\frac{\lambda}{\lambda_0} - 1 = \frac{v}{c} \Rightarrow \frac{\lambda - \lambda_0}{\lambda_0} = \frac{v}{c} = z .$$

This gives [1.10].

– Sirius' radial velocity:

Using [1.2], we obtain:

$$v_r = \left(1 - \frac{f}{f_0}\right)c \qquad\qquad [1.11]$$

Knowing that $f = c/\lambda$, we obtain from [1.11]:

$$v_r = \left(1 - \frac{\lambda_0}{\lambda}\right)c \qquad\qquad [1.12]$$

Numerical Application (NA):

– $\lambda_0 = 656.279$ nm (in the laboratory reference frame at rest);

– $\lambda = 656.263$ nm (in the frame of reference linked to the center of inertia of the star Sirius, moving at speed v_r).

$$v_r = \left(1 - \frac{656.279}{656.263}\right) \times 300000000 = -7{,}314 \text{ m·s}^{-1} = -7.3 \text{ km·s}^{-1}$$

The result is in excellent agreement with the -7.6 km·s^{-1} value accepted today. The minus sign means that the star Sirius was approaching us when we measured the Balmer alpha line (H$_\alpha$) at wavelength 656.263 nm. Recall the erroneous value found by Huggins in 1871: -40 km/s.

Note that relation [1.10] leads to the same result ($c = 300,000$ km·s^{-1}):

$$v = \frac{\lambda - \lambda_0}{\lambda_0} c \Rightarrow v = \frac{656.263 - 656.279}{656.279} \times 300000 = -7.3 \text{ km} \cdot \text{s}^{-1}.$$

Note that in practice, the redshift of a galaxy is assessed on the galaxy spectrum by measuring the difference between the wavelength of the hydrogen emission line λ measured (position of the first peak; the second peak is due to nitrogen) and the wavelength known in the laboratory: $\lambda_0 = 6562.8$ Å (Figure 1.4).

NGC 3627-Hydrogen α (6562.8 Å)

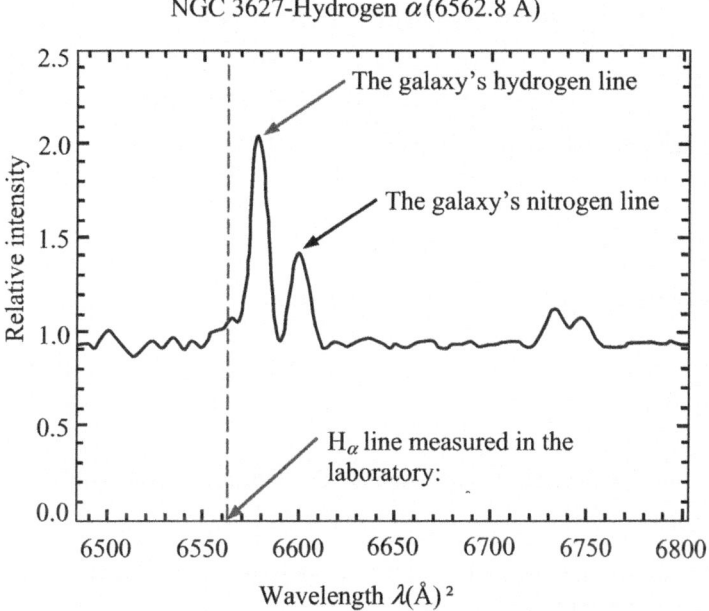

Figure 1.4. *Illustration of the principle of redshift measurement using the spectrum of the galaxy named NGC 3627 (NGC: New General Catalogue). The galaxy's redshift is evaluated on the galaxy's spectrum by measuring the difference between the wavelength of the hydrogen emission line λ measured (position of the first peak, the second peak is due to nitrogen) and the wavelength known in the laboratory: $\lambda_0 = 6562.8$ Å. For a color version of this figure, see www.iste.co.uk/sakho2/ nuclearphysics2.zip*

Using [1.10], we determine the velocity v of the galaxy's distance, i.e.:

$$z = \frac{(\lambda_{measured} - \lambda_0)}{\lambda_0} = \frac{v}{c} \Rightarrow v = \frac{(\lambda_{measured} - \lambda_0)}{\lambda_0} c \,. \qquad [1.13]$$

Here $\lambda_0 = 6562.8$ Å, $c = 300{,}000$ km·s^{-1} and v in km·s^{-1}.

APPLICATION 1.3.–

The measured redshift of galaxy NGC 3627 is $z = 0.002425^*$.

Deduce its distance velocity and measured wavelength *(see: https://archive. wikiwix.com/cache/index2, 2019).

ANSWER.–

Using [1.13], we find: $v = 727.5$ km and $\lambda = 6578.71$ Å.

APPLICATION 1.4.–

Show that the wavelength of the H$_\alpha$ line of the Balmer series is equal to 656.28 nm.

Data [SAK 20]:

– *Balmer's empirical law for the hydrogen atom*:

$$\lambda = \lambda_0 \frac{m^2}{m^2 - 4} \qquad [1.14]$$

In this relationship, λ_0 is the value of the limiting wavelength of the series ($m = \infty$), i.e. $\lambda_0 = 364.56$ nm.

For the record, the law [1.14] was discovered in 1885 by Swiss physicist and mathematician Johann Jakob Balmer (1825–1898).

ANSWER.–

In the emission spectrum of the hydrogen atom, the line H$_\alpha$ corresponds to the transition from the upper $m=3$ level to the 2 level. Using [1.14], we find $\lambda = 656.21$ nm # 656.28 nm.

For the record, the wavelengths λ of the Balmer series lines in the visible range are H_α (orange red), H_β (blue green), H_γ (indigo blue) and H_δ (violet). These are located at 656.3 nm, 486.1 nm, 434.0 nm and 410.2 nm respectively. All other lines are in the ultraviolet.

NOTE.–

i) A galaxy is a gravitationally driven collection of stars and their possible planets, gas, interstellar dust, perhaps mostly dark matter, and often containing a supermassive black hole at its center. The supermassive black hole in our Galaxy (the Milky Way) is called Sagittarius A*. It is located at the center of the Milky Way's gravitational potential. Its mass is estimated at approximately $4 \times 10^6 \, M_\oplus$ (4 million solar masses) [SAK 23b].

ii) The New General Catalogue (NGC) is a directory of 7840 celestial objects. These are mainly galaxies, as well as star clusters and nebulae. The NGC was introduced in 1888 by Irish-Danish astronomer John Dreyer (1852–1926).

iii) In 1777, French astronomer Charles Messier (1730–1818) created a catalog of nebulae and star clusters, known as the *Messier catalog* to distinguish comets (a comet is a small celestial body consisting of a nucleus of ice and dust orbiting (unless disturbed) a star) from various diffuse objects whose nature was unknown at the time (galaxies, various types of nebulae and star clusters). In this Messier catalog, the Crab Nebula named M1 is the first object. The NGC Galaxy in the constellation Leo is named M66.

1.2. Theoretical and experimental facts leading to the validation of the Big Bang model

1.2.1. *From redshift observation to the "primitive atom" hypothesis*

– In 1913, American astronomer Vesto Melvin Slipher (1875–1969) first observed the radial velocities of spiral nebulae, revealed by the enormous "redshifts" of absorption lines in their spectra [SLI 13]. He showed that the Andromeda nebula is approaching the solar system at an average speed of 300 km· s^{-1}. He devoted much of his time to this type of study, in order to verify whether all spiral nebulae have such rapid movements [BRÉ 06; BRÉ 08]. Slipher's discovery provided the first proof of the now widely accepted theory of an expanding universe.

– In 1922, the Russian physicist and mathematician Alexander Friedmann (1888–1925) was the first to publish a theory of the expansion of the universe in his popular science book, *Zeitschriftfür Physik*. In the history of cosmology, this work constitutes the first popularized formulation of the concepts of an expanding or contracting universe and cosmic singularity [LUM 97].

– In 1927, Belgian astrophysicist Georges Lemaître (1894–1966) wrote his paper entitled "Un univers homogène de masse constante et de rayon croissant, rendant compte de la vitesse radiale des nébuleuses extragalactiques" (A homogeneous universe of constant mass and increasing radius, accounting for the radial velocity of extragalactic nebulae) [LEM 27]. Lemaitre linked the expansion of the universe derived from the cosmological solutions of general relativity with observations on the escape velocity of extragalactic nebulae. He proposed the "primitive atom" hypothesis to explain the origin of the universe, introducing the notion of instant zero. For him, if the universe is expanding today, it was much denser in the distant past. Revised and corrected over time, this conception became known as the Big Bang theory [LUM 97, CHA 10].

1.2.2. *From Hubble observations to the discovery of the cosmic microwave background*

In 1920, American astronomer Edwin Hubble (1889–1953) began observing the redshift spectrum of a number of galaxies using the Hooker telescope at the Mount Wilson Observatory in California. Hubble observed that the individual points lay roughly on a straight line (Figure 1.5). He then showed that the variation of velocity with distance is linear, according to the law:

$$v = H_0 d. \hspace{4cm} [1.15]$$

This law expresses the fact that the universe is expanding. This suggests that it had an origin that was later called the Big Bang. In this law, H_0 denotes the *Hubble constant* and d is the distance of the considered galaxy. The subscript 0 is used to indicate the value of the constant at the present moment. Hubble's own measurement of his constant gave the value $H_0 = 558$ km.s^{-1}·Mpc^{-1}. This leads to an age of the universe that is too short (2 billion years) compared with the geological estimate of the Earth's age (approximately 3.5 billion years) [CHA 10].

The constant H_0 was first estimated at the Palomar Observatory in California in 1958 by American astronomer Allan Sandage (1926–2010). He found the value $H_0 = 75$ km ·s^{-1}·Mpc^{-1} [BOI 18]. In 2001, this value was estimated at $H_0 = 72 \pm 8$ km s^{-1}·Mpc^{-1} [FRE 01]. Today, calculation models give the value $H_0 = 67.9 \pm 1.5$ km s^{-1}·Mpc^{-1} [CHO 20] or $H_0 = 70$ km ·s^{-1}·Mpc^{-1}[LOM 20]. The latter value of the Hubble constant means that the universe is expanding 70 km per second faster every 3.26 million light-years (since 1 Mpc = 3.263 million light-years).

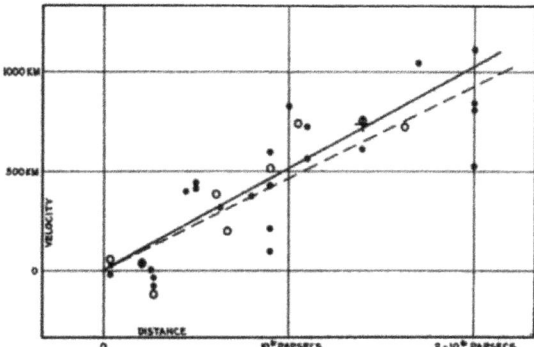

Figure 1.5. *Diagram of Hubble's discovery of the expansion of the universe. The distance–velocity relationship of 46 extragalactic nebulae was studied, and only the individual distances of 24 of the 46 nebulae were estimated [HUB 29]*

In 1927, Lemaitre showed that galaxies move away from each other at a speed approximately proportional to their distance. For this reason, Hubble's law [1.15] was renamed Lemaitre–Hubble's law at the 30th General Assembly of the International Astronomical Union (IAU), held August 20–31, 2018 in Vienna, Austria.

NOTE.–

i) The parsec (contraction of "parallax-second"), symbol pc, is a unit of length used in astronomy. It is worth exactly $648,000/\pi$ *astronomical unit* (AU), or about 3.263 light-years. The astronomical unit is equal to the distance between the Earth and the Sun, and is worth exactly 1 AU = 149,597,870,700 m. The light-year (*l.y.*) is the distance traveled by light in a vacuum at the speed of 2,997,924,58 $m \cdot s^{-1}$ during one Julian year (365.25 days or 31,557,600 seconds), i.e.: 1 *l.y.* = 9,460,730,472 × 10^{15} m.

ii) The Big Bang model thus rejects the second cosmological model based on *steady-state theory* proposed in the late 1940s by British astronomers Fred Hoyle (1915–2001), Austrian astronomer Thomas Gold (1920–2004) and Austro-British astronomer Hermann Bondi (1919–2005). This *Hoyle–Gold–Bondi cosmological model* assumes that the universe is eternal and unchanging, and is unable to account for many astrophysical observations, such as the existence of the CMB at 2.7 K, the abundance of chemical elements in the universe (notably He: the composition of all stars is close to H (80%), He (18%) and other elements 2%), the end of stars (as in the case of the Sun), the end of the universe, etc.

iii) Hoyle, with his strong personality, had little sympathy for his scientific opponents. At a meeting in Pasadena in 1960, he mocked Lemaître, greeting him with the words: "This is the Big Bang man." However, Hoyle's contribution to the Big Bang theory was twofold: firstly, by giving it its name (which for him was a sign of derision), and more seriously, by helping to resolve the question of the abundance of chemical elements in the universe [LUM 97].

In 1965, the discovery of the *cosmic microwave background* (CMB) by American physicists Arno Penzias (b. 1933) and Robert Wilson (b. 1936) provided the decisive argument in favor of the Big Bang theory. The CMB is a highly homogeneous microwave radiation [λ = 30 cm (1 GHz) to 1 mm (300 GHz)] at 2.7 K that has been bathing the universe for approximately 13.8 billion years, cooling as it expands. The existence of the CMB was predicted in 1948 by naturalized American Russian physicist and astronomer Georges Gamow (1904–1968) and American cosmologists Ralph Asher Alpher (1921–2007) and Robert Herman (1914–1997). Working on a hot Big Bang model, they put forward the hypothesis of a primordial universe composed solely of neutrons (which would subsequently decay into protons). Using matter density estimates of the time, they estimated the temperature of the CMB at 5 K [CHA 10]. Therefore, from 1965 onwards, cosmologists believed that the universe was born from a great explosion (Big Bang) of the primitive Lemaitre atom, some 13.8 billion years ago (see the next section).

The WMAP (*Wilkinson Microwave Anisotropy Probe*) and Planck satellite missions (WMAP launched on June 30, 2001, Planck launched in 2009) set out to determine the values of the various cosmological parameters (*see note*), as well as the value of the Hubble constant. To this end, the WMAP mission used the location of the first acoustic peak in the power spectrum of the cosmic microwave background to determine the size of the universe at the time of recombination (*decoupling surface*). The time taken for light to travel to this surface (*see note*) makes it possible to establish a fairly precise age for the universe. WMAP has determined an age of 13.772 ± 0.059 billion years [BOI 18]. In addition, the age of the universe can be determined from the Hubble constant. The best estimates today give an age equal to 13.77 billion ± 40 million years [CHO 20], or approximately 13.8 billion years since the Big Bang.

Similarly, the Planck mission analyzed the anisotropies (temperature variations) of the cosmic microwave background. Thanks to its two optical instruments, HFI (High Frequency Instrument) and LFI (Low Frequency Instrument), the satellite launched in 2009 was able to precisely map the cosmic microwave background. Precise constraints were placed on cosmological parameters, and a value of 67.8 km·s^{-1}·Mpc^{-1} was determined for the Hubble constant. Combined with other external data, this enabled us to calculate an age of the universe equal to 13.799 ± 0.021 billion years [BOI 18].

Let us summarize the origin of the FDC.

At a temperature of 3,000 K, nuclei and electrons combine to form the first atoms; this period in the universe is called recombination. The primordial fluid then passes from the ionized state to the neutral state until reionization. This transition does not alter the spectrum of the background radiation, because at the time of recombination, the number of photons is 109 per nucleon. However, the thermodynamic equilibrium between photons and matter is broken. Photons decouple from matter and move without interaction over long distances. The universe becomes transparent to radiation: this is the emission of the cosmic microwave background [SAZ 13].

In addition, it is important to clarify the notion of the last diffusion surface before continuing with the presentation.

By definition, the surface of last diffusion is the region of space from which the FDC was emitted. It is a spherical region centered on the observer, the furthest from the universe currently accessible to observations. In the first moments of the extremely dense and hot universe, energy is in the form of immense electromagnetic radiation. However, the universe remains opaque to this radiation so that any photon emitted is almost immediately absorbed by the surrounding matter. The vast majority of photons interact mainly with electrons circulating freely in the universe. As it expands, the universe cools and becomes less dense up to a temperature of 3,000 K, allowing free electrons to begin to combine with atomic nuclei to form the first atoms. At this moment, the photon–matter interaction via free electrons disappears very suddenly. The universe then suddenly becomes transparent to light; this is what we call recombination.

Just before recombination, photon-free electron interactions take place by Thomson scattering (scattering of a low-energy photon onto a charged particle of matter at rest, generally a free electron). During recombination, these incessant interactions suddenly cease due to lack of free electrons. The photons then begin to move in a straight line. The fact that the photons interacted with the free electrons, or diffused one last time at the time of recombination, justifies the term "last diffusion". Therefore, the surface of last diffusion is the surface where the last free photon–electron interactions were produced. However, it is important to note that the final scattering surface does not have the particular properties of a surface like that of a star, such as the Sun. Rather, it is the set of points in the observable universe from which the photons of the cosmic microwave background were emitted during recombination. These points are located at such a distance that it took light about

13.7 billion years to reach us by traveling in a straight line. As the universe is homogeneous and isotropic, these points are all located at equal distance from the observer: the surface of last diffusion is therefore a sphere whose center is the observer.

The distance that separates us today from the surface of last diffusion is approximately 43 billion light-years, or more than three times the distance that the light of the cosmic microwave background traveled between its time of emission and now. It was only 40 million light-years away at the time the radiation was emitted. The ratio between these two distances gives the value of the redshift that the cosmic microwave background underwent between its emission and reception: approximately 1,100 (43,000/40 = 1,075). This means that distances were 1,100 times smaller at the time, and the universe was about a billion times $(1,100^3)$ denser than it is today. The temperature of the cosmic microwave background was 1,100 times higher than its current value (2.7 K), or approximately 3,000 K (see Application 1.5). The fact that the cosmic microwave background took 13.7 billion years to travel a distance which was initially 40 million light-years comes from the fact that at the time of its emission, a photon from the cosmic microwave background was traveling in a straight line, but in the interval in which it moved forward a kilometer, the region we are in had moved away by well over a kilometer due to the expansion of the universe. Only when the expansion of the universe sufficiently slows down can photons finally get closer to their destination point.

NOTE.–

The universe most likely contains four types of matter: photons, neutrinos, atoms and molecules constituting baryonic matter, and an unknown form of matter called dark matter (probably made up of as yet undetected elementary particles). Cosmological parameters are quantities involved in the description of cosmological models. The latter are used to correctly describe the observable universe by adjusting them to account for all cosmological observations. Let us cite some cosmological parameters (their current values are given in parentheses: the Hubble constant H0 (current value of the Hubble parameter H: $H_0 = 67.9 \pm 1.5$ km·s^{-1}· Mpc^{-1}), the parameter noted Ω_m linked to the matter, including classical matter (Ω_b, "b" put for baryonic) and dark matter (Ω_{DM}, "DM" put for dark matter): $\Omega_m = \Omega_b + \Omega_{DM}$ ($\Omega_b = 0.03$; $\Omega_{DM} = 0.27$; $\Omega_m = 0.3$), the related parameter to radiation Ω_r (which is today negligible, worth less than 0.01% of the energy content of the universe, then $\Omega_r < 10^{-4}$), the parameter linked to the cosmological constant, also called dark energy Ω_Λ ($\Omega_\Lambda = 0.7$; the latter therefore represents nearly 70% of the mass-energy content of the universe), the cosmological parameter is linked to the curvature of space–time Ω_k, etc.

APPLICATION 1.5.–

At the last diffusion, the photons follow a Planck law. After the radiation–matter decoupling, they only undergo the expansion of the universe, which leads to a dilution of their temperature, according to [SAZ 13]:

$$T(z) = T_0(1 + z). \hspace{4cm} [1.16]$$

In this relation, T_0 is the current temperature and $T(z)$ is the temperature at a spectral shift z.

Show that the temperature of the cosmic microwave background was higher than its current value (2.7 K), approximately 3,000 K.

ANSWER.–

According to the above, the temperature of the cosmic microwave background was 1,100 times higher than its current value $T_0 = 2.7$ K. Thus, for a spectral shift $z = 1,100$, equation [1.1] gives:

$$T(1100) = 2.7\,(1 + 1100) = 2972.7\ K \approx 3000\ K. \hspace{2cm} [1.17]$$

1.3. Brief description of the chronology of the universe's evolution after the Big Bang

The chronology of the universe's evolution after the Big Bang is illustrated in Figure 1.6.

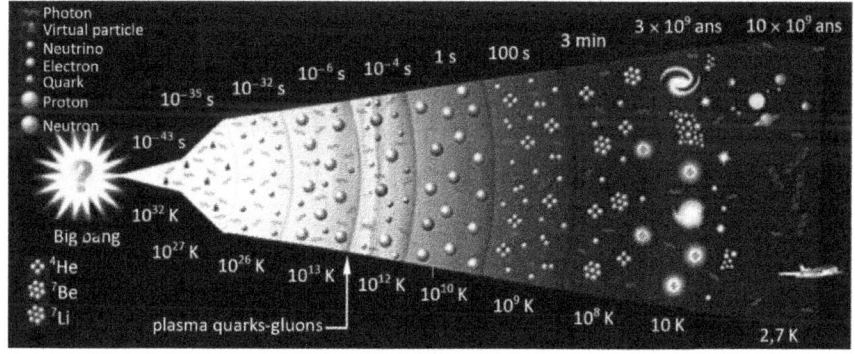

Figure 1.6. *Chronology of the universe's evolution after the Big Bang. For a color version of this figure, see www.iste.co.uk/sakho2/nuclearphysics2.zip*

This chronology is characterized by several eras relating to the evolution of the universe beyond the singularity.

1.3.1. *From singularity to the era of inflation*

The period from singularity to inflation is marked by three eras:

1) Era of Planck ($t < 10^{-43}$ s): during this era, the *four fundamental interactions* of the universe (the electromagnetic interaction, the weak interaction, the strong nuclear interaction and the gravitational interaction) are theoretically unified into a single "theory of everything". These four interactions are described as follows:

– The *electromagnetic interaction*: interaction between charged particles by exchange of photons (*boson vectors* of mass $m = 0$) associated with the electromagnetic interaction (infinite range).

– The *electroweak* or *weak nuclear interaction*: it is responsible for *lepton* interaction (electron, neutrino, etc.) and beta radioactivity. The carrier bosons are the W (W$^+$ and W$^-$) and Z^0 particles, with masses of 86 m_p and 97 m_p respectively (m_p is the mass of the proton). The short-range weak nuclear interaction acts within nucleons. The W and Z bosons are unstable and have very short lifetimes, of the order of 10^{-25} s. It is therefore impossible to observe a W or Z boson directly before it decays: we observe it indirectly by observing the particles that remain after their decay.

– The attractive *strong interaction*: it takes place through the exchange of bosons called "gluons". The name "gluon" comes from "glue", allowing quarks to be "glued" together by the strong interaction. The mass and electric charge of a gluon are zero. The nucleon-bonding interaction is a consequence of the strong interaction: it is responsible for the cohesion of atomic nuclei. Nucleons in the nucleus exchange mesons or pions π^0, π^+ or π^- with masses 134.976 6 MeV/c^2 and 139.570 1 MeV/c^2 (π^+ and π^-) respectively. It is based on the *quantum field theory* of quarks and gluons.

– *The gravitational interaction* (the weakest): it is exerted attractively between different masses (infinite range). It takes place through the exchange of hypothetical bosons called gravitons.

However, singularity poses a serious problem.

Figure 1.6 shows a question mark affirming that our universe was indeed born at an "instant 0" predicted by Lemaître and now known as the Big Bang or *initial singularity*. The existence of a singularity in the Big Bang model is troubling from an intellectual point of view. The reason is quite simple: we cannot speak of an initial instant, since time and space do not exist at this "moment". This is the great weakness of the Big Bang model. What happened before? Nobody has the answer. The following is a summary of our current research into the resolution of the singularity, part of which was presented at our recent conference at the Université Gaston Berger de Saint Louis [SAK 23a].

Before the Big Bang, our evolutionary space and time did not exist. The laws of physics did not exist either. If there really was a Big Bang, i.e. a great explosion, it must have happened somewhere, and the resulting noise is made up of primitive sound waves that propagated in a universe that is not our own, i.e. the one born after the Big Bang.

Logic would dictate that we postulate the existence of an *eternal primitive universe* U_0, before the Big Bang, where Lemaitre's primitive atom exploded. U_0 is an eternal, infinite universe in which all events are stationary: the notions of age (temporal evolution) and death do not exist. A living being or any other object has a fixed age and therefore lives forever. However, space in universe U_0 is elastic. This justifies the possibility of the expansion of our universe in U_0 since the Big Bang. We can also imagine that we are not the only civilizations in universe U_0. There are probably other Big Bangs that have created other universes in U_0. We put forward the fundamental hypothesis that primitive balls of gaseous dark matter were placed at different points O (x_0, y_0, z_0) in the universe U_0 (*seven balls are chosen by way of illustration*). This dark matter, which absorbs all radiation and visible matter, is the opposite of our ordinary matter, which we call gray matter (like black and gray bodies). Each ball was animated by its own rotational movement (analogous to the Earth's rotation around the pole axis). Then, dark energy was injected into one of the targeted balls (Figure 1.7).

The gaseous dark matter instantly ignited. The result was high radiation pressure at a temperature in excess of 10^{32} K. Under the effect of this pressure, the primitive ball exploded at time $t_0 = 0$ in the universe U_0. The result was the expansion of a new universe (our universe) from this initial instant. Our universe is therefore expanding within the eternal, unchanging universe U_0. Other universes can logically be created within universe U_0, given its infinite extent. These different universes are likely to be populated by beings with intelligence superior or inferior to our own.

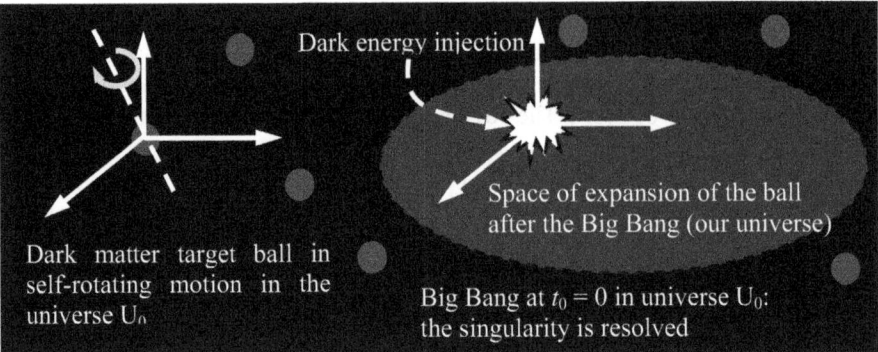

Figure 1.7. *Existence of a universe U0 where the Big Bang occurred at time t0 = 0. There is therefore no singularity. Our Big Bang universe has been expanding ever since into the eternal primitive universe U_0 [SAK 23a]. For a color version of this figure, see www.iste.co.uk/sakho2/nuclearphysics2.zip*

Today, scientists are asking the question: Why do stars revolve around themselves? Why do planets revolve around the Sun and why does our Galaxy revolve around itself? Here is one possible answer (Daniel Pfenniger, see: http://obswww.unige.ch›).

This raises the fundamental problem of the *origin of rotation*. At first sight, if the *initial universe has no rotation of its own*, and every piece of the universe is purely expanding, *with no local rotation*, we *should not get any rotating stars!* In fact, what happens is that, as the universe expands, gravitational instability is triggered at a critical moment. This instability is at the origin of all the stars and structures we know: when gravity becomes stronger than thermal pressure (which tends to homogenize), the tendency to collapse in on itself dominates. Thus, small initial fluctuations in matter density are greatly amplified over time. The same applies to small fluctuations in rotation. "Vortices" of all sizes form and amplify spontaneously, rather like eddies in a river. The total amount of rotation remains very small indeed, but virtually all stars rotate! We can say that for every star rotating in one direction, there is another rotating in the opposite direction, so that the total rotation cancels out.

However, there is a problem with this answer, which we find very interesting: What is the origin of gravity? Without it, small fluctuations in rotation and "vortices" of all sizes could not form. Our universe U_0 model sheds light on the very origin of gravity and the rotation of all bodies in our universe.

We hypothesize that the ball's initial own rotational kinetic energy was transferred to all matter formed during the era of grand unification, at which point the universe was rich in photons, quarks, electrons, neutrinos and *virtual particles*. This transfer was mainly in the form of rotational and translational kinetic energy (by virtue of the principle of energy conservation). This explains the origin of gravity and, in turn, the rotational motion of all the stars in our universe (including the rotational motion of electrons in atoms).

Moreover, this transfer of energy in the form of kinetic energy explains the phenomenon of galaxy recession (escape). However, the expansion of the universe is explained by considering that dark matter (m_d) is permanently injected into our universe to "inflate" it, like a balloon into which air is injected. This "inflating" of the universe justifies its expansion since the Big Bang. The expansion will stop when no more dark matter is injected. The balance of gravitational forces will then break down, and stars and galaxies will scatter in all directions, marking the beginning of the end of the universe. In the recent past, we have postulated a fundamental equation giving the power radiated by an accelerating star in any gravitational field [SAK 16b]. This equation enables the radius of black holes to be calculated. It is also used to calculate the evaporation time of planets and stars within the framework of the gravitational collapse model of rotation centers in solar and galactic systems.

The analogy of the death of our universe can be made with that of the Sun. The Sun consumes hydrogen via a fusion reaction at a temperature of 15 million Kelvin to produce helium-4. The depletion of this nuclear fuel will trigger the end of the Sun (Figure 2.3, Chapter 2). By analogy, our universe feeds on the dark matter with which it is filled. When this matter is no longer injected into our universe from outside, its expansion will subside, triggering the beginning of the end of the world. Finally, let us note that dark energy E_d and dark matter satisfy the mass-energy equivalence relation $E_d = mc_d$ (conversion of dark energy into dark matter and vice versa). This assumes the existence of "non-luminous" dark photons moving in the universe U_0 at the speed $c_d > c$ (speed of the luminous photon).

Note that none of our current laws can explain the properties of dark matter in the universe U_0. The reason is simple: our laws only govern the behavior of ordinary matter in the universe born after the Big Bang. We therefore need to invent new laws of physics that will incorporate the existence of dark photons propagating at the speed of $c_d > c$.

2) The grand unification era (10^{-43} s at 10^{-35} s): during this era, gravity separates from the other three interactions, which remain unified (Figure 1.8). The unification of the three interactions is made possible by the energy-scale dependence of their *coupling constants*. A coupling constant is a number describing the intensity of an interaction. In quantum field theory, the coupling constant is directly proportional to energy.

At very high energies, the different values of the interaction coupling constants converge to the same value. This justifies the possibility of their unification [BOI 18]. During the era of grand unification, the universe is rich in photons, quarks, electrons, neutrinos and *virtual particles*.

In practice, the coupling constants of the strong interaction (α_S), the electromagnetic interaction (α_{EM}) and the weak interaction (α_W) are measured in particle accelerators. The calculation of their evolutions, thanks to the normalization group equation (differential equation), seems to converge at an energy of 10^{15} GeV, the so-called *grand unification mass*. The evolution of the $1/\alpha$ inverses, which are close to being straight lines, have no known point of intersection [VET 23] (Figure 1.8).

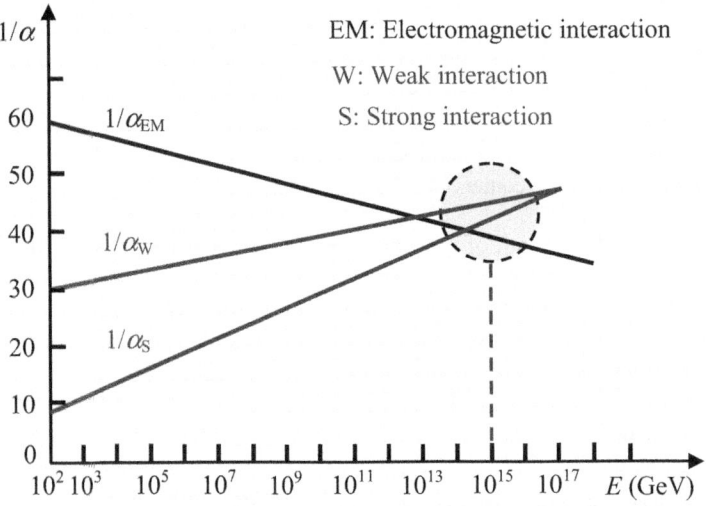

Figure 1.8. *Variation of the inverse of the coupling constants of the strong (α_S) electromagnetic (α_{EM}) and weak (α_W) interactions with particle energy. The evolution of the inverses 1/α that are close to being straight lines have no known point of intersection. The straight lines appear to converge at an energy of 1015 GeV. For a color version of this figure, see www.iste.co.uk/sakho2/nuclearphysics2.zip*

NOTE.–

Electroweak theory is the unified description of the weak interaction and the electromagnetic interaction. The theory predicts the masses of the gauge bosons (or vector bosons) of the weak interaction to be 86 m_p for the $W^{\pm}$ bosons and 97 m_p for the Z^0 boson. The photon, the vector boson of the electromagnetic interaction, has zero mass. These mass differences explain the considerable difference in the behavior of these interactions at low energies. The electroweak theory is the Glashow–Weinberg–Salam theory of the 1960s, derived from the major and largely independent contributions of American physicists Sheldon Lee Glashow (b. 1932) and Steven Weinberg (1933–2021), and Pakistani Muhammad Abdus Salam (1926–1996). This contribution earned them the 1979 Nobel Prize in Physics. In particle physics, a gauge boson is an elementary particle of the boson class (particle with integer spin) acting as the carrier of an elementary interaction. Interacting elementary particles exert forces on each other by exchanging gauge bosons, usually in the form of virtual particles.

3) The inflation era (10^{-35} s at 10^{-33} s): during this period, the universe underwent a very rapid, homogeneous and isotropic expansion phase, in an extremely short time at temperature $T = 10^{26}$ K (1,000 million billion billion kelvin). It grew by a factor of at least 10^{26}.

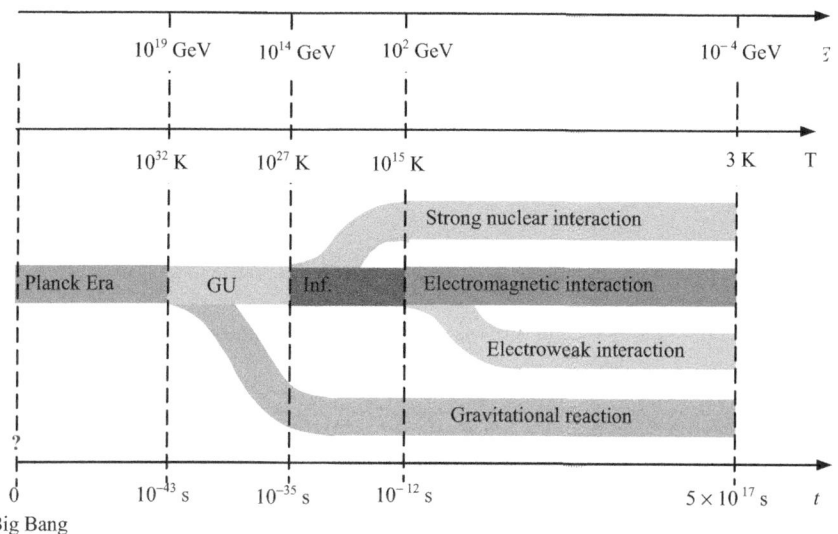

E: particle energy, T: temperature of the universe, t: moment of the universe.

Figure 1.9. *Illustration of the grand unification (GU): separation of the gravitational force from the other three forces at 10^{-43} s at the end of the Planck era. During the inflationary era (inf.) between 10^{-35} s and 10^{-33} s, the strong nuclear interaction separates from the electromagnetic and weak nuclear interactions, which remain unified until 10^{-12} s. At t 10^{-12} s until today, the four forces or interactions are completely separated at the cosmic microwave background temperature of 3 K (2.7 K). For a color version of this figure, see www.iste.co.uk/sakho2/ nuclearphysics2.zip*

During inflation, part of the photon energy becomes virtual quarks and virtual hyperons. Hyperons are heavier than nucleons and made up of a combination of three quarks, at least one s (strange) quark: for example, u [up, + (2/3)e], d [down, -(1/3)e], s (-1/3)e. Ex: the hyperon lambda Λ (uds) with mass 1115.6 MeV/c^2. From 10^{-35} s, the strong nuclear force separates from the electromagnetic force and the weak nuclear force (electroweak force), which remain unified until 10^{-12} s (Figure 1.9).

The unification of the strong interaction and the weak nuclear interaction presupposes the existence of a particle called the *Higgs boson*. The end of inflation thus marks the appearance of matter and antimatter in the universe: quarks, photon, electron, neutrino and their antiparticles – the photon is its own antiparticle.

NOTE.–

The Higgs boson is an elementary particle that is thought to give non-zero mass to certain gauge bosons, such as the W and Z bosons of the electroweak interaction. The particle is named after British physicist Peter Higgs (b. 1929), who proposed its existence in 1964, independently of Belgian physicists François Englert (b. 1932) and Robert Brout (1928–2011), American physicists Gerald Guralnik (1936–2014) and Carl Richard Hagen (b. 1937), and British physicist Tom Kibble (1932–2016), who named it the *massive scalar boson*.

1.3.2. *From baryogenesis to primordial nucleosynthesis*

Between 10^{-32} s and 1 s, the universe's first nucleons and nuclei appeared during three main eras.

1) The baryogenesis era (10^{-32} s to 10^{-12} s): this was the era of the formation of *baryons,* heavy particles made up of three quarks, the best-known of which are the proton (*uud*) and neutron (*udd*). At this time, the temperature of the universe was approximately 10^{13} K (10,000 billion Kelvin). At this point, the *asymmetry between matter and antimatter* became apparent, with the formation of matter taking precedence over the annihilation process. This preponderance of matter was necessary; otherwise, stars and galaxies could not form.

2) The era of quark–gluon plasma formation (10^{-12} s to 10^{-6} s): during this era, a type of plasma is formed that only exists at extremely high temperatures of 2×10^{12} K (2,000 billion Kelvin). At this temperature, quarks and gluons escape the confinement of the strong interaction and become almost free to move. When the temperature of the universe drops to 10^{10} K (10 billion Kelvin), the plasma disappears and quarks give rise to additional nucleons (protons and neutrons) up to

1 s after the Big Bang. Note that at 10^{-12} s until today, the four interactions are completely separated at the 3 K temperature of the cosmic microwave background (Figure 1.9).

3) The era of nucleosynthesis (1 s to 3 min): this is the era during which neutrons and protons came together at a temperature of 1 billion Kelvin to form the first nuclei. All heavier elements were subsequently formed during stellar and explosive nucleosynthesis processes in stars after their formation 1 billion years after the Big Bang. The different nucleosynthesis processes (primordial, stellar and explosive) are discussed in detail in Chapter 2.

1.3.3. *From the dark age of the universe to the radiative era*

– *The dark age of the universe* refers to the era beginning with the *decoupling of radiation and matter* between three minutes and 380,000 years after the Big Bang, during which the universe contained no stars. During this period, hydrogen and helium nuclei began to attract free electrons and transform into atoms. From 300,000 years onwards, the density of the universe is such that it became opaque to light: this is known as the "*dark age*" of the universe. The only light present was that of fossil radiation. This age coincided with the recombination period.

– *The radiative era extends from 300,000 years to 1 billion years*: after 400,000 years, the temperature of the universe cooled to 10^5 K (100,000 Kelvin), and hydrogen atoms began to ionize. Photons were then free to propagate. The universe was now dominated by the *radiative era*. Photons are the main source of energy in the universe.

NOTE.–

At the same time, photons were now too low-energy to be absorbed by atoms. Light no longer had any obstacles, and photons could now propagate unhindered. The universe became transparent to light, and radiation was said to decouple from matter. This decoupling left a trace that can still be observed today. Since it cannot be absorbed, this so-called fossil radiation still fills the universe. However, its temperature has been divided by a factor of 1,000, as the size of the universe has multiplied by the same factor since the time of recombination. Since the temperature was approximately 3,000 degrees then, the present-day universe must be bathed in radiation at approximately three degrees from absolute zero (2,725 to be precise). According to Wien's law, its maximum is therefore to be found at a wavelength of the order of a millimeter in the microwave range (see demonstration below). See: https://astronomes.com/big-bang/recombinaison-rayonnement-fossile/, 2022.

As mentioned above, the CMB at 2.7 K is microwave radiation in the wavelength range $\lambda = 30$ cm to 1 mm. Using Wien's law formula [2.6a] (equation in Chapter 2), the wavelength of fossil radiation at 2.7 K is:

$$\lambda_{max} = \sigma_w / T \Rightarrow \lambda_{max} = 2{,}898 \times 10^{-3} / 2.7 = 1.07 \times 10^{-3} \text{ m} \approx 1.1 \text{ mm}.$$

This is consistent with microwave fossil radiation.

1.3.4. Star formation

The formation of stars and galaxies took place between 1 billion years ago and today. The first heavy nuclei were created, then the first dust clouds, and finally the first stars and large structures, such as galaxies and galaxy clusters.

A star is a luminous ball of gas, composed mainly of hydrogen and helium. The core of a star reaches an extremely high temperature (several million degrees). This high temperature enables the hydrogen nuclei to fuse together, sustaining the luminous energy that allows us to see them from so far away, for up to tens of billions of years.

The gas making up stars is "ionized", meaning that negatively-charged electrons are totally or partially separated from positively-charged nuclei. This gas is called plasma. With the naked eye or visible light telescopes, we can only see the luminous surface of stars. Thanks to scientific telescopes on land and in space, we can observe the entire electromagnetic spectrum of the star. Each area of the spectrum provides specific information on the origin, evolution and functioning of stars. For example, infrared rays tell us where and how stars form, visible light tells us about their chemical composition, and radio waves tell us about their magnetic activity. Finally, X-rays and gamma rays reveal the very high temperatures reached during explosions at the end of a star's life. The data from all these sources can be used to precisely measure the amount of energy produced by a star, its surface temperature, chemical composition and its impact on the interstellar environment.

The equilibrium of stars in the universe is governed by two opposing effects: gravitation, which tends to compress the gas that makes up the star, and the thermal pressure of this gas, which tends to expand it. The core of a star is extremely hot. The temperature difference between the star's core and surface causes heat, and therefore energy, to flow from the center outwards. This heat is ultimately radiated by the star's surface, which causes it to glow.

The energy of stars comes from the nuclear fusion reactions that take place at their centers. Stars evolve by transforming hydrogen into heavier elements. At the end of their lives, nuclear reactions spiral out of control and stars swell. Eventually, as their internal resources dry up, most of them contract permanently into a very dense star. The star then slowly cools down until it barely shines at all. A star's lifespan thus varies according to its mass. The larger the star, the faster it consumes its energy. A massive star can live for a few tens of millions of years, while smaller stars can live for tens or even hundreds of billions of years [COM 17].

For all stars, "ignition" begins with the initial contraction of a cloud of dihydrogen (H_2) gas, heating up the star's core:

– If the gas mass $M<0.08\ M_\odot$ (0.08 solar mass) [SCA 11], thermonuclear reactions will not ignite. The stellar gas degenerates into a *brown dwarf*, also known as a "failed star". Brown dwarfs are cosmic objects more massive than most planets (between 13 and 75 times the mass of Jupiter), but not heavy enough to become a star [ROS 21]. The fate of such a failed star is eternal cooling, and it does not end up as a *white dwarf* (see note).

– If the mass of gas $M \geq 0.08\ M_\odot$, then the ignition of *hydrogen fusion* occurs. This leads to the formation of a real star. This fusion requires a temperature of 10 million kelvin, and is involved in the nuclear evolution of 90% of the 200 billion stars in our Galaxy [SCA 11]. Moreover, hydrogen fusion can take place in two ways: either via the proton–proton cycle (p–p cycle) in the case of medium-sized stars like the Sun (Figure 2.2, Chapter 2), or via the CNO cycle (Figure 2.14, Chapter 2) in the case of a massive star. It should be noted that the slow fusion of hydrogen justifies the very long lifespan of stars like the Sun, which has already lived half its estimated 10-billion-year lifespan (see Figure 2.5, Chapter 2).

Let us summarize the star formation mechanism [COM 17].

Inside galaxies are gigantic clouds of gas and dust, known as molecular clouds. Their shape has recently been revealed in infrared light. Far from being spherical, they are structured into a network of interstellar filaments. Probably under the combined effect of turbulence and gravitation, some of these filaments may condense, contract and then fragment into pockets of gas: *protostellar cores*. In these cores, the mass of gas gradually accumulates, particle agitation increases and the gas temperature rises to approximately 1 million degrees. At this stage, the star is considered a protostar. Each *protostar* then becomes progressively denser, collapsing in on itself under the influence of gravity. The temperature continues to rise until it is high enough to produce thermonuclear fusion reactions. With these reactions, the protostar becomes a star. Star formation can be observed by astrophysicists, thanks to the infrared and submillimeter radiation emitted by the gas and dust making up protostellar cores.

During their lifetime, the internal structure of stars evolves, as do their rotation rate and magnetic activity. These two processes are intimately linked via an effect known as *fluid dynamo*, which converts mechanical energy into electrical energy. The turbulent and convective zones inside a star, where energy is transported not by light but by the large-scale motion of matter, generate electrical currents. In turn, these currents, combined with the star's rotation, generate magnetic fields via the fluid dynamo effect, whose intensity and structure change over time. In the case of the Sun, for example, the famous 11-year cycle and the periodic appearance of sunspots illustrate the magnetic activity of stars.

What is more, the core of a star at the end of its life is essentially made up of carbon and oxygen. With nuclear reactions no longer occurring, the star loses power and is no longer able to perform its task. The star begins to collapse under its own weight, decreasing in size and increasing in density. Due to the compression of matter, each electron is confined to a tiny space and its position is consequently very well defined. The electrons are thus animated by very rapid movements, and their agitation gives rise to a new type of pressure force, of purely quantum origin, called *degeneracy pressure*. This counteracts the star's collapse and restores equilibrium with the star's gravitational force, which has now turned it into a white dwarf. Due to the high compression of matter, a white dwarf is much smaller and denser than a normal star (*the density of a white dwarf reaches the phenomenal value of approximately 1 ton per cubic centimeter of matter*). The small size of a white dwarf is responsible for its very low luminosity. This is why white dwarfs form a separate group in the *Hertzsprung–Russel diagram* (Figure 1.10), below the main sequence. Once a star has become a white dwarf, its life is marked by only a few minor changes. Since the star no longer has a source of energy, its temperature and luminosity drop. Its color changes from white to red and, after a few billion years, it emits very little in the visible range. It then becomes a *black dwarf*.

For the record, it was in 1910 that Danish chemist and astronomer Ejnar Hertzsprung (1873–1967) and American astronomer Henry Norris Russell (1877–1957) came up with the idea of classifying stars by plotting their *luminosity* (or absolute magnitude) against their *effective temperature*. They obtained a luminosity–temperature graph called the *Hertzsprung–Russell diagram* (abbreviated to *H–R diagram*), enabling them to study stellar populations and establish the theory of stellar evolution. They discovered the *main sequence*, giants, supergiants and dwarfs. The main sequence is the band in the Hertzsprung–Russell diagram where 80% of stars are located. Most of these stars spend the greater part of their lives on the main sequence. Those outside the main sequence are either at the beginning or end of their lives. Each star moves along the H–R diagram. At the end of its life, it leaves the main sequence to become a *giant star* and then a white dwarf.

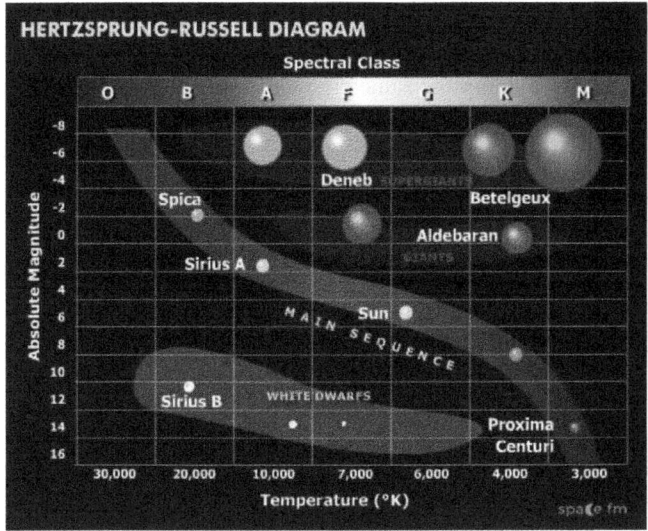

Figure 1.10. *Hertzsprung–Russell diagram. Source: https://www.space.fm/ astronomy/starsgalaxies/hrdiagram.html. Note that the Sun is located on the main sequence. To remember the order of the letters O B A F G K M of a spectral class or type, see Source: https://fr.wikipedia.org/wiki/Diagramme_de_Hertzsprung-Russell, 2021. Note the special case of the stars Sirius A and Sirius B. Source: https:// www.astronomes.com/soleil-etoile/naine-blanche/, 2019. For a color version of this figure, see www.iste.co.uk/sakho2/nuclearphysics2.zip*

NOTE.–

i) The effective temperature of an object, for example, a star, is the temperature of a black body whose surface would emit the same power per unit area as the star; T is therefore the minimum temperature in the upper part of the star's atmosphere.

ii) The luminosity L (*avoid confusion with the luminance L of an emitting body*), the most important characteristic of a star, is the total amount of energy emitted per second at all wavelengths. The relationship between luminosity and the effective or surface temperature T of a star is established using Stefan–Boltzmann's law. Recall that this law expresses the total energy emittance M^0 or the power radiated per unit area by the black body: $M^0 = \sigma T^4$ (σ, Stefan–Boltzmann constant) (see Application 1.2, Chapter 1). Therefore, for an emissive stellar surface S, the luminosity: $L = SM^0 = \sigma S T^4$. Knowing the maximum wavelength of the peak in the star's emission spectrum, we determine its effective temperature T using Wien's first law [2.6a] (see note at the end of Application 2.2, Chapter 2). This then allows us to determine its luminosity L.

Note that astronomers are careful to distinguish between the star's luminosity (total energy production) and the amount of energy that reaches our eyes or a telescope on Earth. Stars emit the same amount of energy in all directions in space. Consequently, only a tiny fraction of the energy emitted by a star actually reaches an observer on Earth. The *apparent luminosity* of a star is therefore defined as the amount of radiated energy that reaches a given surface (say, 1 square meter) on Earth every second. Looking at the night sky, we can see a wide range of apparent luminosities among stars that are so dark that a telescope is needed to detect them.

The Nucleosynthesis Process

Overall objective	
Describe the processes of primordial, stellar and explosive nucleosynthesis that led to the formation of all the chemical elements known in the universe.	
Specific objectives	
Define nucleosynthesis;	Understand the chemical composition of stars at the end of stellar nucleosynthesis;
Learn about the different nucleosynthesis processes;	Write the photodegradation reaction of iron-56 leading to the formation of a neutron star;
Define spallation and photodisintegration;	Describe the neutron capture processes (s processes) that led to the synthesis of nuclei heavier than iron-56, up to lead;
Define primordial nucleosynthesis;	Explain the natural abundance curve of chemical elements formed during the three processes of primordial, stellar and explosive nucleosynthesis;
Learn about the different fusion reactions during primordial nucleosynthesis;	Know the decay mode of elements tending towards the iron peak;
Describe the proton–proton cycle in the case of the Sun;	Describe the triple-alpha reaction;
Understand the hydrogen and helium composition of the universe at the end of primordial nucleosynthesis;	Define the Hoyle state for carbon;
Define stellar nucleosynthesis;	Describe the processes involved in the formation of compound nuclei;
Define the various r, s, p and rp processes involved in stellar nucleosynthesis;	Describe the CNO (Carbon–Nitrogen–Oxygen) or Bethe–Weizsäcker cycle;
Understand the fusion reactions of helium-4, carbon-12, the photodisintegration reaction of neon-20 and the fusion reactions of oxygen-16 during stellar nucleosynthesis;	Know the usefulness of the reaction $^{14}N\ (p,\ \gamma)^{15}O$;
Describe the silicon fusion process;	Define NORM (Naturally Occurring Radioactive Materials) radionuclides;

Know the photodecay products of silicon nuclei;	Define TENORM (Technologically Enhanced Naturally Occurring Radioactive Materials) radionuclides;
Know the role played by silicon 28 in the formation of all elements up to iron;	Distinguish between natural and artificial radionuclides;
Define explosive nucleosynthesis;	Learn about the applications of radionuclides in various areas of everyday life, particularly in archaeology (dating) and nuclear medicine imaging.
Prerequisites	
Chemical element;	Properties of spontaneous nuclear reactions (decay α & β);
Symbol of a nuclide;	Properties of induced nuclear reactions (fission and fusion);
Isotope properties;	Nuclear de-excitation process.

2.1. Nucleosynthesis

According to the *Standard Model*, nucleosynthesis began around 1 s after the *Big Bang* and continued until around 3 min. During this era, at a temperature of 1 billion Kelvin, neutrons and protons grouped together to form the first nuclei. Subsequently, all the elements from carbon to the heaviest elements were formed during the processes of stellar nucleosynthesis and explosive star formation. The first stars were formed a few hundred million years after the Big Bang (see Chapter 1).

The Standard Model of Cosmology is a cosmological model that currently provides the most satisfactory description of the universe's observable history and current composition, as revealed by astronomical observations. The Standard Model describes the universe as a homogeneous, isotropic expanding space. Note that a cosmological model is a mathematical description of all (or part) of the history of the universe. It is based on a theory of physics, usually general relativity or possibly another relativistic theory of gravitation.

2.1.1. *Notion of chemical elements*

At the age of 15, while working as a schoolmaster in Manchester, John Dalton (1766–1844) founded "atomic theory". His first table covered six elements: hydrogen, nitrogen, carbon, oxygen, phosphorus and sulfur. In 1808, he published A New System of Chemical Philosophy, in which he listed the atomic weights of a number of known elements in relation to the atomic weight of hydrogen [SAK 11].

Taking up Dalton's idea, Swedish chemist Jakob Berzelius (1799–1848) suggested using one or two letters to represent chemical elements. Berzelius wondered why not symbolize each body with a letter and write the chemical reaction as an algebraic equation?

For example, if the letter H stands for hydrogen and O for oxygen, the synthesis of water is written:

$$2H + O = H_2O. \hspace{3cm} [2.1]$$

This is how chemists subsequently designated the symbols of all known elements. They either used the first uppercase letter of the element's name, as in the case of hydrogen (H), carbon (C), boron (B), etc., or followed the first uppercase letter with the second lowercase letter of the element's name. This is the case, for example, with helium (He), lithium (Li), neon (Ne), etc. The symbols for some elements depart from this general rule. This is the case with the element Na (from the German "natrium") for sodium, the element Cd for cadmium, the element As for arsenic, the element W (from the German wolfram) for tungsten, etc.

A *chemical element* is made up of the set of entities (atoms or ions) that have the same number Z of protons in their nuclei. Today, there are 118 different chemical elements, 90 of which occur naturally on Earth. The various chemical elements are classified in the periodic table of elements devised in 1869 by Russian chemist Dimitri Mendeleyev (1834–1907). All the chemical elements known in the universe were created by successive nuclear reactions. For a complete understanding of Mendeleyev's table, we need to go back to the Big Bang (see Chapter 1 for more details).

2.1.2. Definition, different nucleosynthesis processes

Gamow used nuclear physics data developed for military projects to propose a theory of the origin of the elements. He hypothesized that all elements were produced during the first, very hot phases of the expanding universe. Thanks to the work of Hoyle and his collaborators, this view was later corrected: only the lightest isotopes – hydrogen, helium, lithium – could have been formed in the early, hot universe, according to the process contemplated by Gamow [LUM 97]. The heavier elements, such as carbon, nitrogen, oxygen, etc., came from stars created later.

At the time of the Big Bang, a physical phenomenon called *nucleosynthesis* took place. It is defined as the set of physical processes leading to the synthesis of atomic nuclei by nuclear fission or fusion. There are three main ways in which all known nuclei in the universe are formed [BEC 10-11, MAR 10, SCA 11, TAT 15, COM 15, FAR 17, COL 19, ESS 19]:

– primordial nucleosynthesis;

– stellar nucleosynthesis;

– explosive nucleosynthesis.

In addition to these three nucleus-forming pathways, there is *spallation* and *photodisintegration*. Spallation is the formation or destruction of large nuclei by very high-energy particles (mainly protons), and photodisintegration is the destruction of nuclei by gamma photons.

2.1.3. *Primordial nucleosynthesis*

Primordial nucleosynthesis (BBN for "Big Bang nucleosynthesis") describes the formation of light elements such as deuterium, helium-3, helium-4 and lithium-7 in the first instants of the universe (between around 1 and 100 seconds after the Big Bang). Heavier elements are produced in nuclear reactions within stars [BAI 08].

One second after the Big Bang, the temperature of the nascent universe reached 10 billion Kelvin. The universe was then made up of a soup of particles: photons γ, electrons, protons and neutrons. With the *annihilation* of electrons and positrons (*antielectrons*), the photon-dominated universe was more luminous than material: the era of *electromagnetic radiation* began. When the universe was about 100 s old, the temperature, still falling, reached 1 billion degrees (10^9 K). At this point, primordial nucleosynthesis began, with the formation of atomic nuclei from protons and neutrons that had been free since their creation 10^{-8}s after the explosion.

Primordial nucleosynthesis favors the synthesis of light nuclei from the very first 100 seconds of the universe. These are mainly helium (^{3}He and ^{4}He), deuterium (^{2}H) or (D), tritium ^{3}H and lithium (^{7}Li) [BEC 10]. The fusion reactions leading to the formation of these nuclei are as follows [SCA 11]:

$$_1^1H + {}_0^1n \rightarrow {}_1^2H + \gamma \qquad\qquad\qquad [2.2a]$$

$$_1^2H + {}_0^1n \rightarrow {}_1^3H + \gamma \qquad\qquad\qquad [2.2b]$$

$$_1^2H + {}_1^1H \rightarrow {}_2^3He + \gamma \qquad\qquad\qquad [2.2c]$$

$$_1^2H + {}_1^2H \rightarrow {}_2^4He + \gamma \qquad\qquad\qquad [2.2d]$$

$$_2^4H + {}_1^3H \rightarrow {}_3^7Li + \gamma \qquad\qquad\qquad [2.2e]$$

However, these nuclei are unstable as long as the photons present in the area are energetic enough to break the newly created nuclei [2.2]. Fortunately, the temperature of the universe continued to drop, and photons became less and less energetic. This drop in temperature was crucial, as matter creation was important for the future of the universe: the formation of stars, planets, galaxies and the biological matter of which we are made.

As the universe cooled, the temperature dropped to 10^6 K. This made further fusion reactions to nuclei heavier than helium impossible. The chemical composition of the universe remained fixed. In percentage by mass: 75% H (3/4 of the universe) and 25% He (1/4 of the universe). Therefore, during primordial nucleosynthesis, hydrogen and helium predominated.

Note the presence of trace amounts of the light chemical elements deuterium (^{2}H), helium-3 (^{3}He), lithium-7 (^{7}Li) and beryllium-7 (^{7}Be). The network of the most important nuclear reactions in primordial nucleosynthesis is shown in Figure 2.1 [ALA 06].

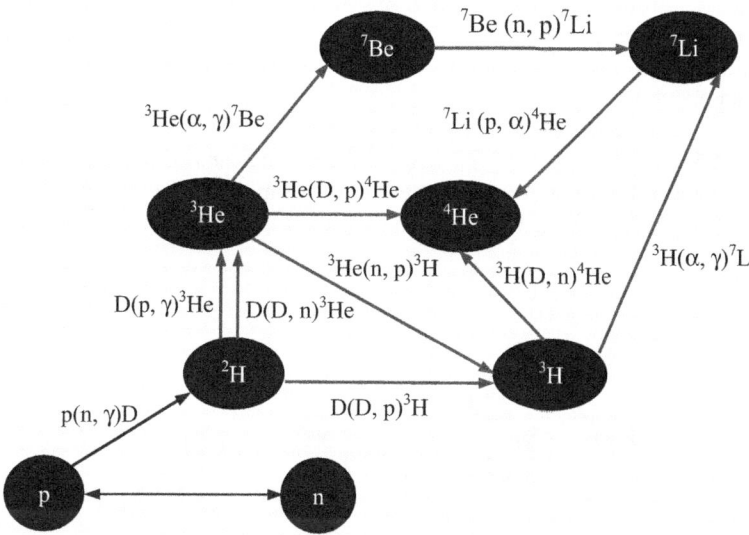

Figure 2.1. *The network of nuclear reactions most relevant to primordial nucleosynthesis. The nuclear physics applied to primordial nucleosynthesis is well mastered in that the nuclear reactions involved can be measured in the laboratory at the energies concerned. In addition to the stable isotopes hydrogen (^{1}H), deuterium (^{2}H or D), helium-3 and 4 (^{3}He,^{4}He), and lithium-7 (^{7}Li), we must also consider the radioactive nuclei of tritium (^{3}H or t) and beryllium-7 (^{7}Be), which decay to ^{3}He in around 12 years and ^{7}Li in around 53 days, respectively. The arrows represent nuclear reactions. Thus, 12 reactions are important for primordial nucleosynthesis. Two of these are evaluated by theory (in blue), the other 10 measured in the laboratory (in red). Note that ^{7}Li is produced in two ways: directly (t + ^{4}He) and via ^{7}Be (^{3}He + ^{4}He) (g stands for a gamma photon, n for a neutron and A for a ^{4}He nucleus, also known as an alpha nucleus) (source: http://books.openedition.org/editionscnrs/docannexe/image/11570/img-1.jpg, 2006). For a color version of this figure, see www.iste.co.uk/sakho2/nuclearphysics2.zip*

Note that the nuclei ^{3}H (half-life $T = 12$ a) and ^{7}Be (half-life $T = 53$ j) decay, respectively, by radioactivity β^- and by electron capture, and will therefore eventually disappear from the BBN products. The corresponding equations are as follows:

$$^3_1H \rightarrow {}^3_2He + {}^0_{-1}e + \bar{v}_e$$ [2.2f]

$$^7_4Be + {}^0_{-1}e \rightarrow {}^7_3Li + \bar{v}_e + RX$$ [2.2g]

According to the above, primordial nucleosynthesis ends with the creation of hydrogen, helium and a little lithium-7. We will then have to wait several billion years for the chemical composition of the universe to slightly change, when the temperature conditions in stars allow *nuclear reactions* leading to the formation of the heavier elements.

Let us take a look at the special case of thermonuclear fusion in the Sun's core ($T = 15$ million Kelvin). At the center of the Sun, four ^{1}H protons fuse in a *proton–proton cycle* to form a ^{4}He helium nucleus (Figure 2.2).

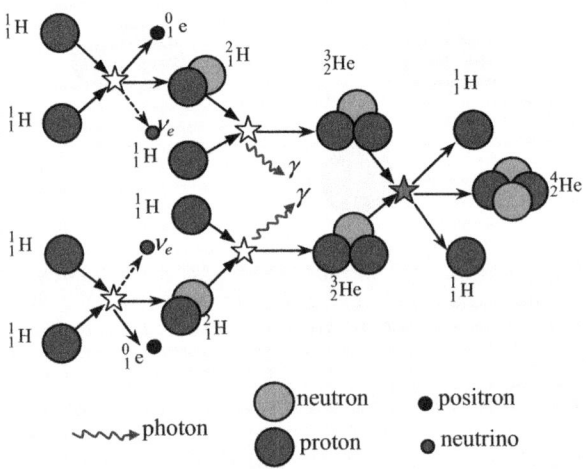

Figure 2.2. *Proton–proton cycle for the Sun. For a color version of this figure, see www.iste.co.uk/sakho2/nuclearphysics2.zip*

As shown in Figure 2.2, the proton–proton cycle comprises two half-cycles whose processes are absolutely identical. For each half-cycle, two protons fuse to give a ^{2}H *deuteron*, an v_e *electron neutrino* and a *positron* (β^+ or ^0_1e):

$$2\,^1_1H \rightarrow\ ^2_1H +\ ^0_1e + v_e \qquad\qquad [2.3a]$$

Subsequently, each deuteron formed fuses with a proton to form a helium-3 nucleus and a gamma photon. This puts an end to the half-cycle:

$$^2_1H +\ ^1_1H \rightarrow\ ^3_2He + \gamma \qquad\qquad [2.3b]$$

Finally, the two helium-3 nuclei formed fuse to form two protons and a helium-4 nucleus, marking the end of the proton–proton cycle:

$$^3_2He +\ ^3_2He \rightarrow\ ^4_2He + 2\,^1_1H \qquad\qquad [2.3c]$$

Summing up equations [2.3], the overall equation of the two cycles translating thermonuclear fusion in the solar core is written (based on Figure 2.2):

$$4\,^1_1H \rightarrow\ ^4_2He + 2\gamma + 2\,^0_1e + 2v_e \qquad\qquad [2.3d]$$

The Sun, made up of hydrogen (75% by mass or 92% by volume) and helium (25% by mass or 8% by volume), fuses 700 million tons of hydrogen nuclei into 695 million tons of helium nuclei every second. Every second, *the Sun loses an estimated 4.3 million tons of mass* (see Application 2.2) through *thermonuclear fusion*, which provides the Sun with energy that is radiated throughout space.

In the general case, there are three series of nuclear fusion reactions leading to the formation of helium-4 at the heart of stars. These series of reactions, known as *proton–proton chains*, are listed in Table 2.1.

Chain pp1	Chain pp2	Chain pp3
p(p, β^+v)d	p(p, β^+v)d	p(p, β^+v)d
d(p, γ)^{3}He	d(p, γ)^{3}He	d(p, γ)^{3}He
^{3}He(^{3}He, 2p)α	^{3}He(α, γ)^{7}Be	^{3}He(α, γ)^{7}Be
	^{7}Be(β^-, v)^{7}Li	^{7}Be(p, γ)^{8}B
	^{7}Li(p, α)α	^{8}B(β^+v)^{8}Be
		Be(α)α

Table 2.1. *The three proton–proton chains leading to the formation of helium-4 in the cores of stars. For each chain, a total of four protons fuse to form a helium nucleus, releasing an energy of 26.731 MeV [FRU 18]*

As shown in Table 2.1, each series begins with the fusion of two p protons (^{1}H), leading to the formation of a d deuterium (^{2}H).

APPLICATION 2.1.–

1) Which of the reaction chains listed in Table 1.1 corresponds to the one evolving at the heart of the Sun?

2) Specify the number of hydrogen nuclei converted into a helium nucleus 4. Deduce the energy released Q. Express the result in MeV.

Data: atomic masses: ^{1}H: 1.008142 u; ^{4}He: 4.003873 u; 1 u = 931.5 MeV/c^2.

ANSWER.–

1) For the Sun, the pp1 chain corresponds to the fusion reaction. However, this is the first cycle (Figure 2.2). If we consider both cycles, we find the balance of reactions [2.2d]. In total, this corresponds to the fusion of four protons into a helium-4 nucleus via the global equation [2.3d].

2) If we do the counting, four hydrogen nuclei are converted into one helium-4 nucleus (remember that the pp1 chain is a half-chain). The energy released is:

$Q = (4\ m_{1H} - m_{4He})\ c^2 = (4 \times 1.008142 - 4.003873) \times 931.5 = 26.729$ MeV $\approx$ 26.730 MeV.

This corresponds to the release of an energy of 26.731 MeV [FRU 18] (see the legend in Table 2.1).

Note that the energies released in nuclear reactions are generally estimated from atomic masses (and not from the masses of the nuclei involved in these reactions).

During primordial nucleosynthesis, a common reaction known as *nuclear spallation* takes place. This is an X $(a,\ b)$ Y reaction, in which a nucleus X is impacted by an *incident particle a* (neutron n, proton p, etc.) or absorbs high-energy electromagnetic radiation (between 50 MeV and a few GeV). The reaction produces an *emerging particle b* (neutron n, proton p, etc.) and light nuclei Y (deuterium ^{2}H, helium-3 or -4, lithium-7, etc.), according to the equation:

$$a + X \rightarrow b + Y \qquad [2.4]$$

Nuclear spallation is also involved in the interaction of *cosmic ray particles* (see Chapter 3, section 3.1.2 for more details) with matter. This important phenomenon, known as *cosmic spallation*, produces light elements such as lithium and boron by bombarding matter with cosmic ray particles from the Sun or Galaxy. Cosmic rays consist mainly of protons and helium.

APPLICATION 2.2.–

By virtue of *Wien's first law* (see note), the *Sun's surface temperature* is estimated at 5,800 K. Show that the mass loss of the Sun per second is approximately equal to 4.3×10^{9} kg. Let us assimilate the Sun to a spherical *black body* of radius $R = 696,342,000$ m.

Data: *Stefan–Boltzmann law* (expressing the *total energy emittance M^0 or the power radiated per unit area by the black body*): $M^0 = \sigma T^4$; $\sigma = 5.66897 \times 10^{-8}$ $W \cdot m^{-2} \cdot K^{-4}$ (*Stefan–Boltzmann constant*): *mass of the Sun*: $M_S = 1.99 \times 10^{30}$ kg; *speed of light in vacuum*: $c = 3.0 \times 10^8 m \cdot s^{-1}$.

ANSWER.–

The solar surface $S = 4\pi R^2$. By virtue of the *mass–energy equivalence relation* (*Einstein's relation*) $\Delta E = \Delta mc^2$. The energy radiated by the solar surface per unit time is therefore equal to:

$$\Delta E = \Delta mc^2 = M^0 \times S \times \Delta t = \sigma T^4 \times S \times \Delta t \qquad [2.5a]$$

Hence, the Sun's mass loss per second is written as:

$$\Delta m = \frac{4\pi\sigma T^4 \times R^2}{c^2} \qquad [2.5b]$$

Numerically, we find:

$$\Delta m = \frac{4\pi \times 5.66897 \times 10^{-8} \times (5800)^4 \times (6.96342 \times 10^8)^2}{(3 \times 10^8)^2} = 4.343388415 \times 10^9 \, kg \cdot s^{-1}$$

or $\Delta m \approx 4.3 \times 10^6$ tons $\cdot s^{-1}$. Remember that Δm does not have the unit of a mass, but rather that of a mass lost per unit of time.

A NOTE ON WIEN'S LAWS.–

The laws giving the spectral distribution of blackbody radiation at short wavelengths were published in 1896 by the German physicist Wilhelm Wien (1864–1928). There are two of these laws, known as Wien's laws [SAK 19].

– *Wien's first law*: the wavelength λ_{max} at the peak of the blackbody isotherm shifts to shorter wavelengths with increasing temperature according to the law:

$$\lambda_{max}T = \sigma_w \qquad\qquad [2.6a]$$

In the law [2.6a], σ_w is *Wien's constant*: $\sigma_w = 2.898 \times 10^{-3}$ m K.

– *Wien's second law*: the ordinate $M^0_{\lambda max}$ of the *monochromatic emittance* maximum is proportional to the fifth power of the temperature, i.e.:

$$M^0_{\lambda max} = BT^5 \qquad\qquad [2.6b]$$

In the law [2.6b]:

– $B = 1.28 \times 10^{-5}$ W $\cdot$m$^{-3}\cdot$ K^{-5} if λ_{max} in m and $M^0_{\lambda max}$ in W $\cdot$m^{-3};

– $B = 1.28 \times 10^{-11}$ W $\cdot$m$^{-3}\cdot\mu$m$^{-1}\cdot$ K^{-5} if λ_{max} in μm and $M^0_{\lambda max}$ in W $\cdot$m$^{-2}\cdot\mu$m^{-1}.

2.1.4. Stellar nucleosynthesis

Stellar nucleosynthesis refers to the nuclear fusion reactions that take place in the cores of stars, resulting in the production of most atomic nuclei [BUR 57].

In the general case, the composition of all stars similar to the Sun is close to H (80%), He (18%) and other elements 2% (there are, however, stars that are very poor in metals, and the abundance of helium also varies with the age of stellar populations).

Relative to the mass $M_\odot$ of the Sun ($M_\odot = 1.989 \times 10^{30}$ kg), the evolution of a contracting star can be represented as a function of its mass, according to the standard model of stellar evolution [AUT 09].

In a given star, a hydrostatic equilibrium is established during the so-called "quiet" phase. This equilibrium is achieved by the opposing effects of gravitation, which tends to compress the star, and thermonuclear reactions at the star's core (energy and high temperature), which oppose this contraction (gas pressure + radiation pressure).

During stellar nucleosynthesis, a number of processes take place: *r, s, p, rp* and α.

– The *r process* (*r* stands for *rapid*) is a *neutron capture reaction* by nuclei occurring at high temperature and high neutron density. This nucleosynthesis process enables the formation of heavy elements from the fusion of very heavy nuclei (supernova explosions).

– The *s process* (*s* stands for *slow*) is a neutron capture reaction by heavy nuclei, leading to the synthesis of heavier nuclei. It occurs at lower temperatures and neutron densities than the *r process* (reactions in the last phases of supergiant stars).

– The *p process* (*p* stands for *proton*) is a *proton capture reaction* by atomic nuclei. This process leads to the formation of heavy nuclei from lighter ones.

– The *rp process* (*rp* stands for *rapid proton capture*) involves a series of successive proton captures by a nucleus. While the *r* and *s* processes lead to the synthesis of neutron-rich nuclei; the *rp* process leads to the synthesis of proton-rich nuclei.

– The α process resulting from the capture of helium-4 nuclei: these reactions begin when silicon 28 melts at temperatures above 3 billion Kelvin and give rise to a set of processes known as the "silicon quasi-equilibrium". During the α process, the silicon 28 core undergoes photodisintegration producing α particles (^{4}He nuclei). These are absorbed by other silicon 28 nuclei and generate nuclei such as sulfur 32, argon 36, calcium 40, etc. (see below, after Figure 2.4).

Furthermore, the evolution of stars during stellar nucleosynthesis occurs in two main stages, depending on the size of the star compared to the mass of the Sun:

– During their lives, stars create a large proportion of the elements between lithium-7 and iron-56.

– At the end of their lives, when massive stars (*supernovae*) explode, *explosive nucleosynthesis* takes place (see below), leading to the formation of most elements heavier than iron-56.

In the general case, the evolution of stars during stellar nucleosynthesis involves a series of contraction phases allowing thermonuclear fusion of the chemical elements within the star, starting with hydrogen. The slow fusion of hydrogen explains the long lifespan of stars. Hydrogen fusion at 10 million Kelvin produces helium-4 [SCA 11].

Figure 2.3 illustrates the Sun's evolution from birth to white dwarf. Its lifetime is estimated at 10 billion years (see Application 2.3).

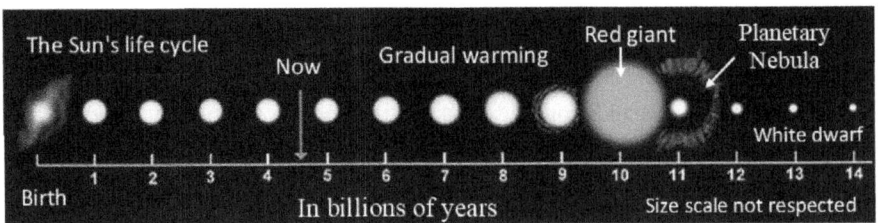

Figure 2.3. *The Sun's evolution over the last 4.5 billion years. Its death is expected in 5 billion years. For a color version of this figure, see www.iste.co.uk/sakho2/nuclearphysics2.zip*

APPLICATION 2.3.–

The Sun, 75% hydrogen by mass, fuses 700 million tons of hydrogen nuclei into 695 million tons of helium nuclei every second. Assuming that only 15% of the fuel burns in the solar core, estimate, in years, the lifetime τ of the Sun whose mass $M_\odot = 1.989 \times 10^{30}$ kg. Take 1 year = 365.25 days.

ANSWER.–

1s $\rightarrow m_\mathrm{H}$ = 700 million tons

$$\tau \rightarrow 15\% \times (75\% \times M_\odot) \tag{2.7}$$

The Sun's lifetime is then:

$$\tau = \frac{15\% \times (75\% \times 1.989 \times 10^{30})}{7 \times 10^{11} \times 365.25 \times 24 \times 3600} = 1.0129 \times 10^{10} \text{ years} \tag{2.8}$$

or $\tau \approx$ 10 billion years.

After exhausting their hydrogen fuel, which has been completely transformed into helium, the star's core contracts again, provided it has sufficient mass. Thus, depending on the mass of the star compared to that of the Sun, two main processes are observed [SCA 11]:

– If the star's mass M is slow ($M < 1/3\ M_\odot$), contraction becomes insufficient to trigger thermonuclear fusion of helium 4. The star then cools down, eventually transforming into a *black dwarf.*

– If the star is massive enough ($M > 1/3\ M_\odot$), contraction causes helium-4 to fuse at a temperature of 100 million Kelvin ($T = 10^8$ K). This fusion creates the elements ^{8}Be (unstable), ^{12}C and ^{16}O, according to equations [2.9]:

$$_2^4He + _2^4He \rightarrow _4^8Be + \gamma \qquad\qquad [2.9a]$$

$$_4^8Be + _2^4He \rightarrow _6^{12}C + \gamma \qquad\qquad [2.9b]$$

$$_6^{12}C + _2^4He \rightarrow _8^{16}O + \gamma \qquad\qquad [2.9c]$$

Nuclear reactions [2.9] show that, after helium-4 depletion, the star's core is composed mainly of carbon-12 (stable) and oxygen-16 (stable), since beryllium-8 is radioactive with a decay period of 6.7×10^{-17} s. It decays by fission, producing two helium-4 nuclei (reverse reaction [2.9a]).

After the helium fuel has been used up, a further contraction of the star's core may first cause the ^{12}C to fuse. Thus [SCA 11]:

– If the star is not massive enough ($1/3\ M_\odot < M < 8\ M_\odot$), contraction will not trigger thermonuclear carbon-12 fusion. The star's fate will be analogous to that of the Sun (Figure 2.3): relapse, transformation into a *red giant* (increasing the star's volume) and then into a *white dwarf*.

– If the star is massive enough ($M > 8\ M_\odot$), contraction causes carbon-12 to fuse at a temperature of 800 million Kelvin ($T = 8 \times 10^8$ K). This fusion produces new elements such as neon-20 and sodium-23, according to equations [2.10]:

$$_6^{12}C + _6^{12}C \rightarrow _{10}^{20}Ne + _2^4He \qquad\qquad [2.10a]$$

$$_6^{12}C + _6^{12}C \rightarrow _{11}^{23}Ne + _1^1H \qquad\qquad [2.10b]$$

If, on the contrary, the temperature rises above 1,100 million Kelvin (1.1×10^9 K), the predominant melting reaction of carbon-12 leads to the formation of magnesium 23:

$$_6^{12}C + _6^{12}C \rightarrow _{12}^{23}Mg + _0^1n \qquad\qquad [2.10c]$$

While neon-20 and sodium-23 are stable, magnesium-23 is radioactive β^+ ($T = 11.317$ s). The formation of stable magnesium-24 requires even higher temperatures. Thus, when it reaches 1,500 million Kelvin (1.5×10^9 K), neon 20 undergoes *photodisintegration*, producing oxygen-16 [2.11a] and fusion with helium-4 to form magnesium-24 [2.11b]:

$$^{20}_{10}Ne + \gamma \rightarrow {}^{16}_{8}O + {}^{4}_{2}He \qquad\qquad\qquad [2.11a]$$

$$^{20}_{10}Ne + {}^{4}_{2}He \rightarrow {}^{24}_{12}Mg + \gamma \qquad\qquad\qquad [2.11b]$$

From 2,500 million Kelvin (2.5×20^9 K), oxygen-26 begins to melt. This produces silicon, sulfur and phosphorus according to the processes [2.12]:

$$^{16}_{8}O + {}^{16}_{8}O \rightarrow {}^{28}_{14}Si + {}^{4}_{2}He \qquad\qquad\qquad [2.12a]$$

$$^{16}_{8}O + {}^{16}_{8}O \rightarrow {}^{30}_{14}Si + 2\,{}^{1}_{1}H \qquad\qquad\qquad [2.12b]$$

$$^{16}_{8}O + {}^{16}_{8}O \rightarrow {}^{31}_{15}P + {}^{1}_{1}H \qquad\qquad\qquad [2.12c]$$

$$^{16}_{8}O + {}^{16}_{8}O \rightarrow {}^{30}_{15}P + {}^{2}_{1}H \qquad\qquad\qquad [2.12d]$$

$$^{16}_{8}O + {}^{16}_{8}O \rightarrow {}^{31}_{16}S + {}^{1}_{0}n \qquad\qquad\qquad [2.12e]$$

After the depletion of oxygen-16, *silicon fusion* ends the fusion process inside the core of massive stars. When the temperature reaches 3,200 million Kelvin (3.2×10^9 K), the silicon nuclei undergo photodisintegration. This produces neutrons, protons and α particles, which in turn interact with the ^{28}Si nuclei present in the medium to form all the elements up to iron-56.

Let us now explain the various fusion reactions involved in the α process. The helium-4 nuclei produced by the photodisintegration of ^{28}Si interact with the remaining ^{28}Si in the stellar medium to form all the elements up to iron-56, according to the following successive reactions:

$$^{28}\text{Si} + {}^{4}\text{He} \rightarrow {}^{32}\text{S} + \gamma \qquad\qquad\qquad [2.13a]$$

$$^{32}\text{S} + {}^{4}\text{He} \rightarrow {}^{36}\text{Ar} + \gamma \qquad\qquad\qquad [2.13b]$$

$$^{36}\text{Ar} + {}^{4}\text{He} \rightarrow {}^{40}\text{Ca} + \gamma \qquad\qquad\qquad [2.13c]$$

$$^{40}\text{Ca} + {}^{4}\text{He} \rightarrow {}^{44}\text{Ti} + \gamma \mapsto {}^{44}\text{Ca} + 2\,\beta^{+} \qquad\qquad\qquad [2.13d]$$

$$^{44}\text{Ti} + {}^{4}\text{He} \rightarrow {}^{48}\text{Cr} + \gamma \mapsto {}^{48}\text{Ti} + 2\,\beta^{+} \qquad\qquad\qquad [2.13e]$$

$$^{48}\text{Cr} + {}^{4}\text{He} \rightarrow {}^{52}\text{Fe} \rightarrow {}^{52}\text{Cr} + 2\ \beta^{+} \qquad\qquad [2.13\text{f}]$$

$$^{52}\text{Fe} + {}^{4}\text{He} \rightarrow {}^{56}\text{Ni} + \gamma \rightarrow {}^{56}\text{Fe} + 2\ \beta^{+} \qquad\qquad [2.13\text{g}]$$

As shown in the reactions [2.13], the formation of iron-56 ends the α process.

Figure 2.4 shows a summary of all the fusion reactions taking place during stellar nucleosynthesis, right up to the formation of iron-56 at the end of the fusion process.

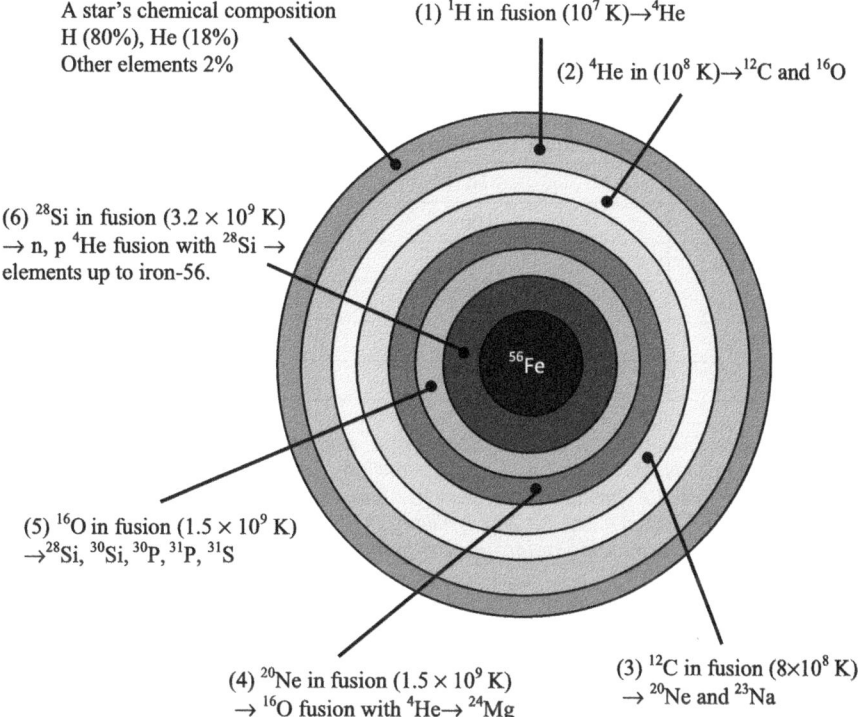

A star's chemical composition
H (80%), He (18%)
Other elements 2%

(1) ^{1}H in fusion (10^{7} K)$\rightarrow{}^{4}$He

(2) ^{4}He in (10^{8} K)$\rightarrow{}^{12}$C and ^{16}O

(6) ^{28}Si in fusion (3.2×10^{9} K) $\rightarrow$ n, p ^{4}He fusion with ^{28}Si $\rightarrow$ elements up to iron-56.

(5) ^{16}O in fusion (1.5×10^{9} K) $\rightarrow{}^{28}$Si, ^{30}Si, ^{30}P, ^{31}P, ^{31}S

(4) ^{20}Ne in fusion (1.5×10^{9} K) $\rightarrow {}^{16}$O fusion with ^{4}He$\rightarrow {}^{24}$Mg

(3) ^{12}C in fusion (8×10^{8} K) $\rightarrow {}^{20}$Ne and ^{23}Na

^{56}Fe

Figure 2.4. *The core of a massive star with an "onion skin" structure before a supernova explosion. In each layer, an element burns and the iron core is inert. For a color version of this figure, see www.iste.co.uk/sakho2/nuclearphysics2.zip*

COMMENTS.–

(1): Hydrogen (^{1}H) melts at 10 million Kelvin: fusion product: ^{4}He.

(2): Helium 4 (^{4}He) melts at 100 million Kelvin: fusion products: ^{12}C and ^{16}O.

(3): Carbon (^{12}C) melts at 800 million Kelvin: products: ^{20}Ne and ^{23}Na.

(4): Neon 20 (^{20}Ne) melts at 1,500 million Kelvin: photodisintegration produces ^{16}O; fusion with ^{4}He produces magnesium-24 (^{24}Mg).

(5): Oxygen (^{16}O) melts at 2,500 million Kelvin: products: ^{28}Si, ^{30}Si, ^{30}P, ^{31}P, ^{31}S.

(6): Silicon (^{28}Si) melts at 3,200 million Kelvin. Photodisintegration of silicon-28 produces neutrons, protons and ^{4}He nuclei. These interact with the remaining ^{28}Si nuclei to form all the elements up to iron-56.

Table 2.2 summarizes the combustibles, fusion temperatures and main nuclei formed during stellar nucleosynthesis [SCA 11, COL 19].

Combustible	Temperature (in millions of Kelvin)	Elements formed
Hydrogen	10	^{4}He
Helium	100	^{12}C, ^{16}N
Carbon	800	^{24}Mg, ^{20}Ne
Neon	1,500	^{16}O, ^{24}Mg
Oxygen	2,500	^{28}Si, ^{30}Si, ^{30}P, ^{31}P, ^{31}S
Silicon	3,200	……………., ^{56}Fe

Table 2.2. *Combustibles, fusion temperatures and nuclei formed during stellar nucleosynthesis. Note that at 1,500 million Kelvin, neon-20 undergoes photodisintegration to produce oxygen-16 and fuses with helium-4 to form magnesium-24. Similarly, at 3,200 million Kelvin, silicon-28 nuclei undergo photodisintegration, producing neutrons, protons and helium-4 nuclei. These particles in turn interact with the ^{28}Si nuclei present in the medium to form all the elements up to iron-56 [2.15]. Fusion in stellar cores stops as soon as iron-56 is formed*

NOTE.–

A *red giant* is a large, bright star that has undergone gravitational collapse of its central part and expansion of its outer parts. A solar-type star evolves into a red giant when the hydrogen in its center is consumed, and the hydrogen in the parts closest to its periphery begins to burn more intensely. Nuclear fusion reactions then begin to consume helium very rapidly. When helium fusion comes to an end, the star starts to contract again. With its low mass unable to reach the temperatures and pressures required to initiate carbon fusion, the core collapses into a *white dwarf* (*an incredibly dense star whose mass is no more than 1.44 times that of the Sun*). The outer layers of the star bounce violently off this solid surface and are thrown into space in the form of a *planetary nebula*. The adjective "planetary" attached to a nebula is linked to it resembling a nebulous-looking disk when observed at low resolution, similar to the appearance of planets. Therefore, the result of this process is a very hot white dwarf surrounded by a cloud of gas composed essentially of hydrogen and helium (and a little carbon) not consumed during fusion.

A *black dwarf* is the hypothetical evolution of a white dwarf star, which has cooled sufficiently to no longer emit visible light. Unlike black holes, which are the final stage in the life of a super-massive star, the black dwarf phase is the fate that awaits medium-sized stars, such as our Sun.

2.1.5. *Explosive nucleosynthesis*

At the end of stellar nucleosynthesis, the star's core is filled with iron-56 (Figure 2.4), the most stable element of all nuclei.

The *Chandrasekhar mass* or *Chandrasekhar limit* is the maximum mass that the electronic degeneracy pressure of a celestial object can support without gravitational collapse. It occurs when matter accumulates around a star made of degenerate matter, such as a white dwarf or the core of a massive star. The mass limit is 1.44 times that of the Sun. Beyond that, the star collapses to form a neutron star or becomes a black hole.

NOTE.–

Indian physicist Subrahmanyan Chandrasekhar (1910–1995) demonstrated in 1930, when he was just 20 years old, that the end of a star depends on its mass and that of its core. He calculated the maximum limit an object can reach before either exploding, in the case of a white dwarf (low-mass star), or collapsing into a neutron star or black hole, in the case of a massive star. This limit is called the "*Chandrasekhar limit*" and is 1.44 solar masses ($1.44M_\odot$) or 2.9×10^{30} kg [BOI 17]. This mass is often approximated to $1.4M_\odot$.

Let M_c be the mass of the core of a star with initial mass M. Two evolutions are possible during a gravitational collapse process:

– If $1.4M_\odot < M_c < 3.3M_\odot$ [AUT 09, SCA 11] corresponding to $8\ M_\odot < M < 30\ M_\odot$, collapse leads to the formation of a neutron star.

– If $M_c > 3.3M_\odot$ or $M > 30\ M_\odot$, the collapsing star turns into a black hole.

Figure 2.5 summarizes the evolution of stars according to the standard model [AUT 09].

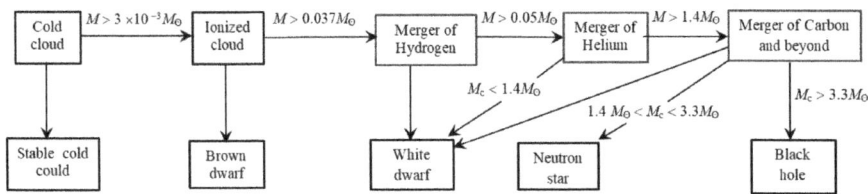

Figure 2.5. *Evolution of a contracting star as a function of mass. Rectangles in blue refer to states of contraction. Rectangles in red indicate merging processes, while those in black indicate the end of stellar life. For a color version of this figure, see www.iste.co.uk/sakho2/nuclearphysics2.zip*

Therefore, for $1.4M_\odot < M_c < 3.3M_\odot$, gravitation prevails and the star's core undergoes a sudden collapse in a matter of seconds. The γ photons radiated into the star's core at tens of billions of Kelvin become energetic enough to cause the photodisintegration of iron-56, according to equation [SCA 11]:

$$\begin{smallmatrix}56\\26\end{smallmatrix}Fe + 26\, \begin{smallmatrix}0\\-1\end{smallmatrix}e + \gamma \;\rightarrow 56\, \begin{smallmatrix}1\\0\end{smallmatrix}n \;+ \nu \qquad\qquad [2.14]$$

As equation [2.14] shows, the photodisintegration of iron-56 leads to the formation of a neutron star.

As the star's core transforms (into a neutron star or black hole), the density of the core increases until it reaches that of atomic nuclei. The star's contraction at extremely high speeds comes to an abrupt halt within fractions of a second. As it falls, the star's outer envelopes bounce off the core, producing a shock wave. This wave sweeps across the star from the center to the outer layers, rekindling fusion in these same layers at billions of Kelvin: a gigantic explosion takes place, known as supernova II (*SN classification: SN IIs have a spectrum containing hydrogen, SN Is have a spectrum containing no hydrogen*).

The supernova disperses all the star's upper layers (enriched in heavy elements) into space. The fast neutrons emitted during this explosion combine with the elements present to form all the chemical elements heavier than iron: this is explosive nucleosynthesis. During explosive nucleosynthesis, all elements heavier than iron are synthesized via two processes: the s process and the r process.

As iron fusion is impossible, the "slow" *neutron capture* process via the s (slow) process is the only nuclear reaction capable of synthesizing elements heavier than iron-56, up to and including lead (see Figure 1.8). Subsequently, a "beta-minus" radioactive decay (in which a neutron is transformed into a proton) generates an element with the next higher atomic number. The general process of neutron capture by a nucleus noted ^{N}X, with N being the number of neutrons, is shown in Figure 2.6 [COL 19].

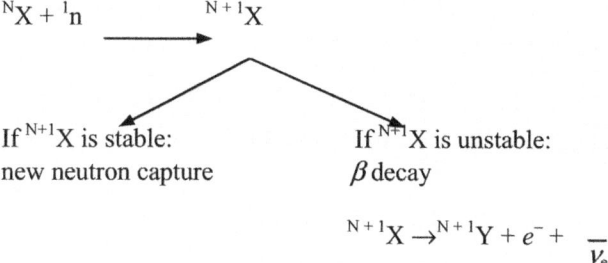

$$^{N}X + {}^{1}n \longrightarrow {}^{N+1}X$$

If ^{N+1}X is stable: If ^{N+1}X is unstable:
new neutron capture β decay

$$^{N+1}X \rightarrow {}^{N+1}Y + e^{-} + \overline{\nu_{e}}$$

Figure 2.6. *Neutron capture processes (s processes)
during explosive nucleosynthesis*

During the *r* process, the flux of neutrons produced during the star's explosion is very high and neutron capture. This is a rapid process (within minutes) of *radiative neutron capture*, followed by β decay (Figure 2.7). Around half of the abundance of elements beyond iron up to uranium is produced by the *r* process.

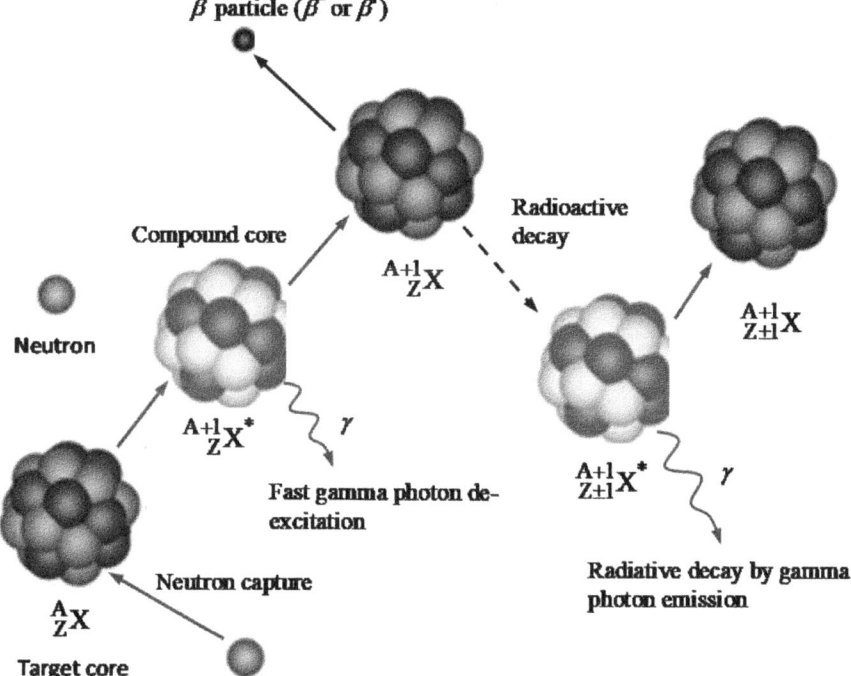

Figure 2.7. *Radiative neutron capture process. For a color version
of this figure, see www.iste.co.uk/sakho2/nuclearphysics2.zip*

As shown in Figure 2.7, after the neutron capture process by the ^{A}X nucleus, the radioactive decay of the compound nucleus $^{A+2}$X yields two products according to the two β decay modes:

$$^{A+1}_{Z}X \rightarrow {}^{A+1}_{Z+1}X + {}^{0}_{-1}e + \gamma + \bar{\nu}_e \ \ (\text{mode } \beta^-) \tag{2.15a}$$

$$^{A+1}_{Z}X \rightarrow {}^{A+1}_{Z-1}X + {}^{0}_{1}e + \gamma + \nu_e \ \ (\text{mode } \beta^+) \tag{2.15b}$$

Figure 2.8 shows the natural abundance of chemical elements formed during the three processes of primordial, stellar and explosive nucleosynthesis.

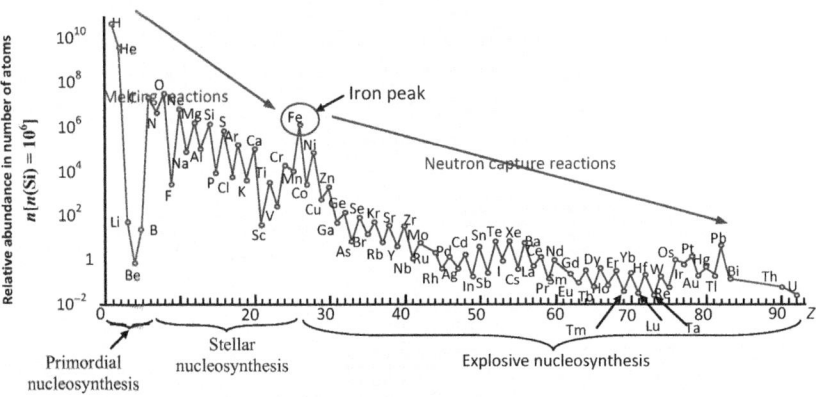

Figure 2.8. *Natural abundance of chemical elements formed during the three nucleosynthesis processes. Note that the relative abundance of elements decreases exponentially with atomic number Z. For a color version of this figure, see www.iste. co.uk/sakho2/nuclearphysics2.zip*

With the exception of lithium, beryllium and boron, all of which show depletion, the abundance of all elements decreases exponentially with atomic number Z. Similarly, elements with even Z atomic numbers are more abundant than their odd-numbered neighbors.

Figure 2.8 also shows a pronounced abundance peak in the vicinity of iron, known as the *iron peak*. As the ^{56}Fe isotope of iron is the most stable element, successive fusions will lead to its formation. Iron fusion is impossible because the energy required to pass the *coulombic barrier* of highly charged nuclei is greater than the energy that can be released by nuclear fusion. This creates a balance between the formation and destruction of iron. Iron's neighbors are the elements chromium (Cr, $Z = 24$), manganese (Mn, $Z = 25$), cobalt (Co, $Z = 27$) and nickel (Ni, $Z = 28$), which will tend towards the iron peak ($Z = 26$) through β decay.

Finally, let us mention the main spallation reactions occurring in the interstellar medium [CHA 15]:

$$\,^4_2He + \,^4_2He \rightarrow \,^6_3Li + \,^1_1H + \,^1_0n \qquad\qquad [2.16a]$$

$$\,^1_1H + \,^{12}_6C \rightarrow \,^{11}_5B + 2\,^1_1H \qquad\qquad [2.16b]$$

$$\,^1_1H + \,^{14}_7N \rightarrow \,^9_4Be + 2\,^1_1H + \,^4_2He \qquad\qquad [2.16c]$$

$$\,^4_2He + \,^{12}_6C \rightarrow \,^7_3Li + \,^1_1H + 2\,^4_2He \qquad\qquad [2.16d]$$

$$\,^1_1H + \,^{14}_7N \rightarrow \,^{10}_5B + 3\,^1_1H + 2\,^1_0n \qquad\qquad [2.16e]$$

VOCABULARY.–

Coulomb barrier: in a description of the interaction of nuclei using energy levels, Coulomb repulsion appears as a barrier called the Coulomb barrier. In the case of nucleosynthesis, the initial state is made up of two nuclei whose meeting must lead to fusion. The higher the atomic number of the nuclei to be fused, the greater the Coulomb barrier and the higher the kinetic energies to be supplied to the initial nuclei. In the cores of stars, temperatures of at least 10^7 K enable such kinetic energies to be generated.

2.2. Other important nucleus-forming processes, radionuclides in the environment

2.2.1. *Triple-alpha reaction, Hoyle state*

We have already seen that the fusion of helium-4 at a temperature of 100 million Kelvin creates, among other things, unstable beryllium-8. The latter fuses with helium-4 to produce carbon-12 via the reaction [2.17b], known as the triple-alpha reaction. In view of its specific nature, we will look at it in detail below.

By definition, the *triple-alpha reaction* refers to a set of nuclear fusion reactions transforming three α particles (^{4}He) into the carbon-12 nucleus via the beryllium-8 nucleus (Figure 2.9).

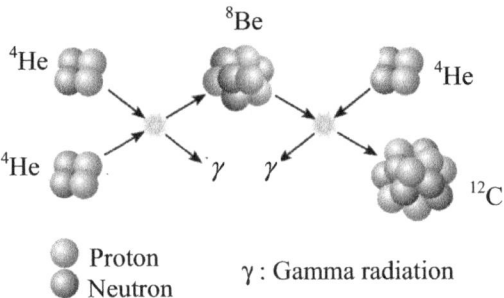

Figure 2.9. *Triple-alpha reaction. The third helium nucleus fuses with a beryllium-8 nucleus to form a carbon-12 nucleus (source: https://en.wikipedia.org/wiki/Triple-alpha_process#/media/File:Triple-Alpha_Process.svg, 2021). For a color version of this figure, see www.iste.co.uk/sakho2/nuclearphysics2.zip*

The triple α reaction plays a very important catalytic role in various explosive stellar processes, where peak temperatures exceed 2 GK [KIR 13], 1 GK = 10^9 K. As a result of the depletion of hydrogen fuel due to a drop in *radiation pressure* in the cores of stars, the latter contract towards a new hydrostatic equilibrium. As the star's core contracts, it heats up to a temperature of between 0.1 GK and 2 GK. This contraction accelerates the fusion of helium-4 nuclei into beryllium-8, according to the equation:

$$^4\text{He} + {}^4\text{He} \rightarrow {}^8\text{Be (ground state)} \qquad\qquad [2.17a]$$

^{8}Be, with a half-life of 8.19×10^{-17} s, decays into two alpha particles. Despite this extremely short half-life, the accelerated fusion of helium-4 via the reaction [2.17a] maintains a sufficient concentration of beryllium-8, which can then fuse with a third helium-4 nucleus (Figure 2.9) to form stable carbon-12, according to equation:

$$^4\text{He} + {}^4\text{Be} \rightarrow {}^{12}\text{C (Hoyle state)} \qquad\qquad [2.17b]$$

In addition, the *Hoyle state* is defined as the second excited state of carbon-12 located at 7654 MeV with an angular momentum-parity (or *spin-parity*) $J^\pi = 0^+$. This state is in resonance with the state of a system of three α particles or with the state composed of an α particle and a beryllium-8 nucleus [SAN 07, FRE 13]. The Hoyle state spontaneously decays to the ground state of ^{12}C, either by passing through its excited state at 4439 MeV and emitting γ photons, or, via an *electron–positron pair creation reaction*, from the Hoyle state directly to the ground state of ^{12}C (Figure 2.10).

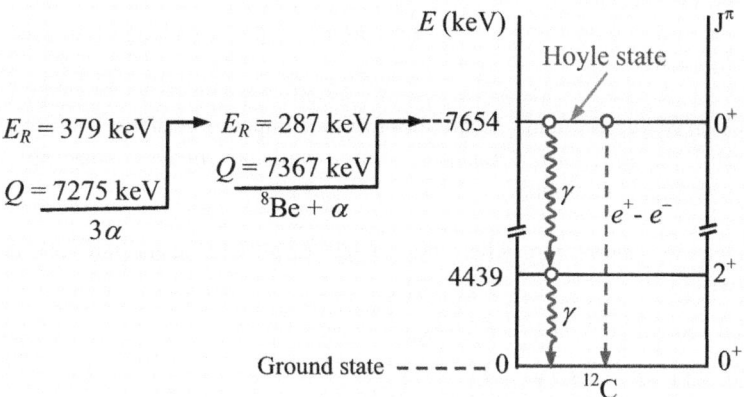

Figure 2.10. *Spontaneous decay from the Hoyle state (located at 7,654 MeV) to the ground state of ^{12}C either by emission of 2 γ photons via the excited level of ^{12}C located at 4,439 MeV or via an electron-positron pair creation reaction. For a color version of this figure, see www.iste.co.uk/sakho2/nuclearphysics2.zip*

2.2.2. *Formation process of compound nuclei, resonance states*

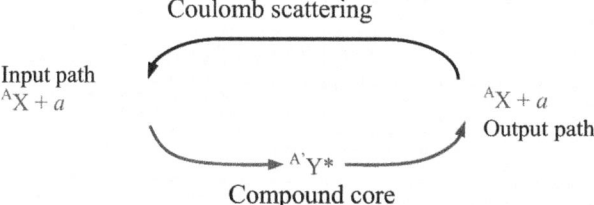

Figure 2.11. *Schematic illustration of the principle of $^{A}X + a$ resonant elastic scattering. In the input channel, we have $^{A}X + a$; in the output channel, we have $^{A}X + a$ (elastic scattering). There are two possible paths from the input to the output path. The first is via Coulomb scattering, with the two nuclei repelling each other and therefore not coming into contact. The second is via the compound nucleus $^{A'}Y^*$, resulting from the fusion of the two nuclei ^{A}X and a (resonant elastic scattering). Subsequently, the compound nucleus decays towards the $^{A}X + a$n output path. For a color version of this figure, see www.iste.co.uk/sakho2/nuclearphysics2.zip*

In the case of *elastic scattering* of a light nucleus *a* (proton, α particle, etc.) by a heavier target nucleus ^{A}X, at least two contributions can be distinguished in the scattering process from the *input path* to the *output path*. These are, on the one hand, the contribution made by the Coulomb potential preventing the two particles from coming into contact and, on the other hand, the contribution made by the fusion of nuclei a and ^AX leading to the formation of the compound nucleus $^{A'}Y*$ (Figure 2.11) [SAN 07].

The fusion of nuclei a and ^AX can only form a compound nucleus $^{A'}Y$ if the latter has discrete energy states. Otherwise, the *compound nucleus* does not form (diffusion remains Coulombian).

Consider the general case where the nuclei are initially in motion. The energy balance gives:

– Initial state: $m_X c^2 + m_a c^2 + E_{cX} + E_{ca}$

– Final state: $m_Y c^2 + E(Y) + E_\gamma$

By virtue of the *principle of conservation of energy*:

$$m_Y c^2 + E(Y) + E_\gamma = m_X c^2 + m_a c^2 + E_{cX} + E_{ca} \qquad [2.18a]$$

Let $E_x = E(Y)$ be the excitation energy of the compound nucleus relative to its ground state. This is a discrete state existing in the compound nucleus $^{A'}Y$. By using [2.18a], we obtain:

$$E_x = (m_X + m_a - m_Y) c^2 + E_{cX} + E_{ca} - E_\gamma \qquad [2.18b]$$

Let Q be the energy due to the mass difference between the initial nuclei and the compound nucleus, and let E_R be the kinetic energy of the nuclei X and a initially in motion. This gives:

$$Q = (m_X + m_a - m_Y) c^2 \qquad [2.18c]$$

$$E_R = E_{cX} + E_{ca} \qquad [2.18d]$$

Therefore, the requirement for the appearance of resonances at certain specific values of the incident kinetic energy E_R is written as follows:

$$E_R = E_x - Q + E_\gamma \qquad [2.19]$$

2.2.3. *CNO (Carbon–Nitrogen–Oxygen) cycle*

The CNO cycle is the main source of energy in stars with masses greater than about 2 solar masses during their hydrogen-burning phase. This cycle is a catalytic sequence of proton capture reactions, followed by β decays on isotopes of C, N and

O. This cycle is a series of nuclear reactions using carbon as a catalyst to fuse four hydrogen atoms into one helium atom with radiated energy (see Figure 2.12).

In general, there are four series of reactions grouped together under the name CNO cycle. These are shown in Table 2.3 [FRU 18].

CNO1	CNO2	CNO3	CNO4
$^{12}C(p,\gamma)^{13}N$	$^{14}N(p,\gamma)^{15}O$	$^{15}N(p,\gamma)^{16}O$	$^{16}O(p,\gamma)^{17}F$
$^{13}N(\beta^+\nu)^{13}C$	$^{15}O(\beta^+\nu)^{15}N$	$^{16}O(p,\gamma)^{17}F$	$^{17}F(\beta^+\nu)^{17}O$
$^{13}C(p,\gamma)^{14}N$	$^{15}N(p,\gamma)^{16}O$	$^{17}F(\beta^+\nu)^{17}O$	$^{17}O(p,\gamma)^{18}F$
$^{14}N(p,\gamma)^{15}O$	$^{16}O(p,\gamma)^{17}F$	$^{17}O(p,\gamma)^{18}F$	$^{18}F(\beta^+\nu)^{18}O$
$^{15}O(\beta^+\nu)^{15}N$	$^{17}F(\beta^+\nu)^{17}O$	$^{18}F(\beta^+\nu)^{18}O$	$^{18}O(p,\gamma)^{19}F$
$^{15}N(p,\alpha)^{12}C$	$^{17}O(p,\alpha)^{14}N$	$^{18}O(p,\alpha)^{15}N$	$^{19}F(p,\alpha)^{16}O$

Table 2.3. *The four series of reactions known as the CNO cycle. As indicated in each column of the table, the result of each series is the fusion of 4 protons into a helium nucleus, producing an energy of 26.731 MeV (see Application 2.1, Answer 2.2)*

It should be noted that the CNO1 cycle is responsible for greater energy production than the pp1 chain (see Table 2.2) for stars with a core temperature greater than 20 MK [FRU 18].

In the case of the Sun, the vast majority of its energy is produced by the pp chain process (see Figure 2.2), with the CNO cycle contributing only around 1%. Consequently, the CNO cycle is only a secondary hydrogen-to-helium conversion mechanism for the Sun. However, detection of the neutrinos it produces can provide direct information on the abundance of the elements C, N and O in the Sun's core. This gives an idea of its metallicity (see Note).

NOTE.–

In general, a stellar medium is characterized by the relative proportion of each of the different elements present. As stated at various points in this section, hydrogen is the majority constituent, followed by helium. The concentration of another element X is defined relative to hydrogen by the ratio of the number of N_X and N_H atoms present in the same volume. By setting the abundance of hydrogen arbitrarily at 12, the "abundance" A of a chemical element X is given by the logarithmic of the ratio N_X/N_H, i.e. $A = \log_{10}(N_X/N_H) + 12$. Solar abundances are used as references for studies of stellar or galactic spectra. These are deduced from the analysis of spectral lines observed in the photosphere, where the corresponding absorptions occur. In general, for a given star, "metallicity" is a number defined as the sum of the mass concentrations of all the star's constituent chemical elements, other than hydrogen or helium. For example, the metallicity of a solar-type star is $Z = 0.02$.

2.2.4. Bethe–Weizsäcker cycle

A particularly well-studied case in astrophysics is the ^{14}N (p, γ)^{15}O reaction involved in the *CNO cycle*. Also known as the *Bethe–Weizsäcker cycle* [MAR 12], this reaction reflects the conversion of hydrogen into helium in massive stars (Figure 2.12).

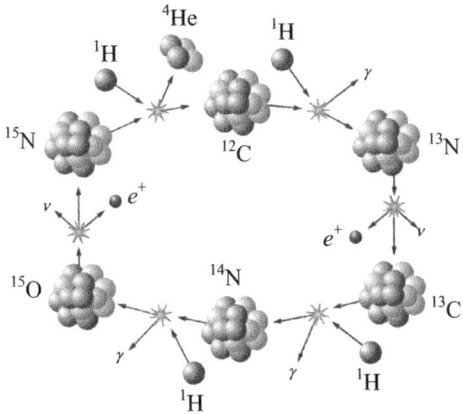

Figure 2.12. *CNO (Carbon–Nitrogen–Oxygen) or Bethe–Weizsäcker cycle. It features the resonant capture reaction ^{14}N (p, γ)^{15}O, leading to the formation of the compound nucleus ^{15}O (source: https://en.wikipedia.org/wiki/CNO_cycle#/media/File: CNO_Cycle.svg, 2021). For a color version of this figure, see www.iste.co.uk/sakho2/ nuclearphysics2.zip*

In the direct capture model, the proton interacts with the whole ^{14}N nucleus, not with its individual nucleons. As the *resonant capture* time required to form a compound nucleus is of the order of 10^{-17}s, the ^{14}N+ p capture reaction is a relatively fast process of the order of 10^{-22}s [DAI 13]. Figure 2.13 shows the special case of the excited levels of compound nucleus ^{15}O obtained by the ^{14}N + p resonant capture reaction [IMB 05, MAR 12, DAI 13].

Q (keV)	E (keV)	E_x (keV)	J^π

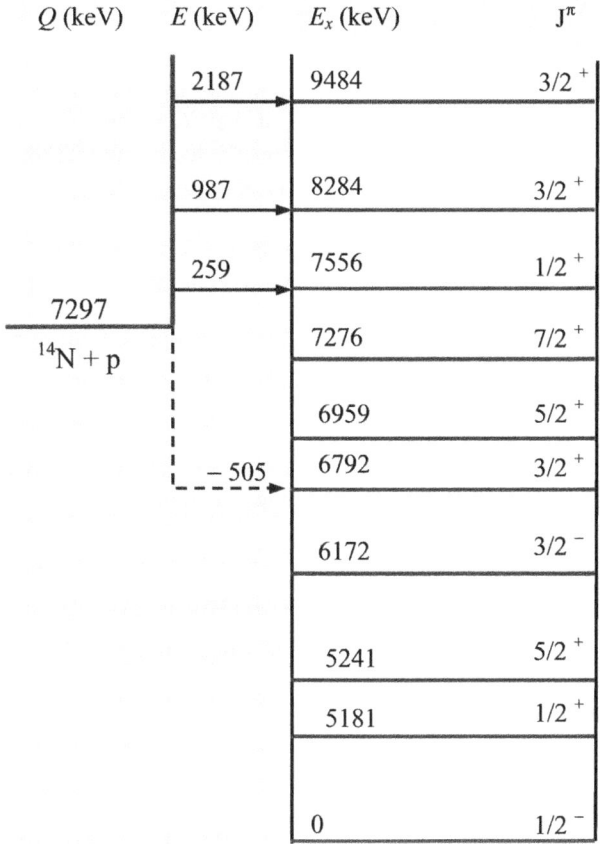

Figure 2.13. *Diagram of the excited levels of the compound nucleus ^{15}O resulting from the nuclear reaction $^{14}N(p, \gamma)^{15}O$. For a color version of this figure, see www.iste.co.uk/sakho2/nuclearphysics2.zip*

Using E_p as the kinetic energy of the proton beam in the laboratory system and E as the effective energy of the beam in the center-of-mass system [MAR 12], the energy of the transition line γ is given by [2.19] (nitrogen nuclei being immobile: $E_{cN} = 0$):

$$E_\gamma = Q + E_p - E_x \qquad\qquad [2.20]$$

Therefore, the direct proton capture reaction ^{14}N $(p, \gamma)^{15}O$ occurs at all incident photon energies E_p. During each transition between the excited levels of the compound nucleus ^{15}O, a photon γ of energy E_γ given by [2.20] is emitted to form a bound state of energy E_x in the final nucleus. For the reaction ^{14}N $(p, \gamma)^{15}O$, the energy due to the mass difference between the initial nuclei and the compound nucleus is $Q = 7.297$ MeV [IMB 05, MAR 12, DAI 13].

For $E_R = 259$ keV, experimental measurements by Imbriani et al. [IMB 05] revealed the existence of four excited states of the ^{15}O compound nucleus. The E_x excitation energies corresponding to these states are located at $(5,180.8 \pm 0.2)$ keV, $(6,172.3 \pm 0.2)$ keV, $(6,791.7 \pm 0.2)$ keV and $(7,556.4 \pm 0.6)$ keV.

At low energies $E < 150$ keV, the experiment reveals the existence of excited states of oxygen ^{15}O, enhancing the contribution of the capture reaction to the ground state. This is the case of the resonance at $E = -505$ keV, corresponding to the 6,792 keV excitation energy of the ^{15}O compound nucleus (Figure 2.13).

APPLICATION 2.4.–

Justify the value of the energy $Q = 7,297$ MeV due to the mass difference between the initial nuclei ^{14}N + p and the compound nucleus ^{15}O.

Data:

– *Atomic masses*: ^{14}N: 14.007515 u; ^{15}O: 15.007768 u; ^{1}H = 1.008142 u.

– *Atomic mass unit*: 1 u = 931.5 MeV/c^2; m_e = 0.00055 u.

ANSWER.–

From [2.2c], we obtain:

$$Q = (m_N + m_p - m_O)\, c^2 = (14.007515 + 1.008142 - 15.007768)\ \text{u} = 7.3486$$
$$\text{MeV} \approx 7,350 \text{ keV}$$

This result is in agreement with the $Q = 7,297$ keV $\approx 7,300$ keV value (see Figure 2.13).

BIOGRAPHY CORNER.–

Hans Albrecht Bethe (1906–2005) was a German-born American physicist. He won the 1967 Nobel Prize in Physics for his contribution to the understanding of stellar nucleosynthesis.

Carl Friedrich (1912–2007) was a German physicist and philosopher. He was the longest-lived member of the research team that attempted to develop atomic weapons in Germany during World War II.

2.2.5. *Natural and artificial radionuclides in the environment*

The radionuclides present in the universe are classified into two categories: NORM (Naturally Occurring Radioactive Materials) and TENORM (Technologically Enhanced Naturally Occurring Radioactive Materials) [SAB 19].

NORMs are *naturally occurring radionuclides*, i.e. those present in the universe without any human intervention. NORMs are made up of precursors and decay products from the radioactive filiations of natural families. NORMs include members of the thorium-232, uranium-235 and uranium-238 radioactive families, as well as the decay products of potassium-40. These radionuclides, which have persisted since the formation of the Earth and the Solar System, some 4.5 billion years ago, are considered to be the main radionuclide sources of external irradiation to the human body. However, NORM remains naturally in the Earth's environment in trace amounts and is therefore of low toxicity. The nuclear radiation produced by NORM decay is known as *telluric radiation*, which occurs naturally in the Earth's crust and atmosphere [SAB 19].

Let us consider the case where *anthropogenic pollution*, resulting from various human activities in the biosphere, leads to high concentrations of naturally occurring radionuclides. In this case, we consider altered radionuclides as anthropogenic or TENORM, according to the United Nations Environmental Protection Agency.

In contrast to natural radionuclides, *artificial radionuclides* refer to all radioactive elements created artificially using a particle accelerator or nuclear reactor. Some artificial radionuclides are used as sources of radiation for various applications (see Chapters 2, 4 and 5). Artificial radionuclides commonly measured in the Earth's environment include strontium-90 (^{90}Sr), iodine-131 (^{131}I), the ^{134}Cs and ^{137}Cs isotopes of cesium, the ^{238}Pu, ^{239}Pu and ^{240}Pu isotopes of plutonium, and americium-241 (^{241}Am). These radionuclides can originate from fallout from the nuclear weapons tests that took place in 1950 and 1960, or from nuclear accidents, such as the Chernobyl nuclear disaster in 1986 and the Fukushima nuclear accident in 2011 [SAB 19].

3

Radiochronometer Applications in Dating

Overall objective	
Understand the properties of some radioelements used in radiochronometry.	
Specific objectives	
Define a cosmogenic isotope;	Describe the CFCS (Constant Flux and Constant Sedimentation), CRS (Constant Rate of Supply) and CIC (Constant Initial Concentration) models;
Distinguish between primary and secondary cosmic rays;	Use age equations derived from CFCS, CRS and CIC models;
Understand the principle and usefulness of carbon-14 dating;	Learn about the effects of nuclear testing and the Chernobyl accident on atmospheric air;
Know the principle of carbon-14 radioactivity detection counters;	Learn about the main radionuclides in fallout from nuclear weapons testing;
Distinguish between radiocarbon age and calendar age (real age);	Distinguish between local fallout, tropospheric fallout and stratospheric fallout;
Know the impact of the "Bomb", "Suess" and reservoir effects on atmospheric ^{14}C content;	Know which radioelements significantly contaminate the consumable parts of plants;
Know the age equation yielding the atmospheric isotope ratio ($^{14}C/^{12}C$) of a sample;	Learn about the environmental impact of radioelements released during the *Chernobyl accident*;
Know the different decay modes of potassium-40;	Know the principle and usefulness of cesium-137 dating;
Know the principle and usefulness of potassium–argon (K–Ar) dating;	Write the nuclear reaction equations leading to the formation of iodine-127 and cesium-137;
Know the accumulation process of argon-40 in a crystallized lava;	Write the cesium-137 decay chain leading to barium-137;

For a color version of all figures in this chapter, see www.iste.co.uk/sakho2/nuclearphysics2.zip.

Establish the age equation for the K–Ar radiochronometer;	Interpret the cesium-137 decay diagram;
Justify the need for the atmospheric correction to the age equation;	Interpret the cesium-137 activity profile as a function of depth in a core sample;
Know how to measure the potassium and argon content of a sample to be dated using the ^{40}K–^{40}Ar clock;	Know the principle and usefulness of beryllium-7 dating;
Define a coring system;	Establish the age equation of a sediment using the beryllium-7 radiochronometer;
Be familiar with the most common methods of radiochronometric dating using a coring system;	Estimate the bioturbation rate D_b or a given sedimentary chronology;
Know the principle and usefulness of lead-210 dating;	Understand the principle and usefulness of uranium–thorium or uranium–lead dating;
Understand the origins of supported lead-210 and excess lead-210 in sediments;	Understand the principle of sampling and mechanical preparation of samples to be dated by the uranium–thorium clock;
Establish the relationship between total lead-210, supported lead-210 and excess lead-210;	Know the principle of coral dating using the ^{238}U/^{230}Th and ^{235}U/^{231}Pa methods;
Determine the secular equilibrium date between supported lead-210 and radium-226;	Interpret time trends in ^{230}Th/^{238}U and ^{231}Pa/^{235}U isotope ratios;
Know the three models for determining the age of a sediment;	Know the principle of coral dating using the ^{234}U/^{230}Th method.
Prerequisites	
Radioactive decay law;	Fission of uranium-238;
γ de-excitation;	Uranium-235 and uranium-238 decay chain;
Law of simple filiation with two bodies;	Natural production of carbon-14;
Law of accumulation of simple filiation with two bodies;	Nuclear spallation reactions;
Properties of β^- decay;	Electron–positron annihilation process.

3.1. Carbon-14 dating

Absolute dating (or absolute radio-chronology) aims to quantitatively estimate the age of rock formations, fossil organisms, climatic events, etc. This method is based on the principle of radioactive decay [CHA 21].

3.1.1. *A brief history of radiocarbon-14 dating*

We give a brief overview of experimental research into the carbon-14 dating method according to Guibert [GUI 18].

Radiocarbon dating is undoubtedly the most popular method in archaeology, and the one that has attracted the most attention worldwide since the 1950s [LIB 49] and Libby's Nobel Prize in Chemistry in 1960. Everyone is familiar with the principle of the method, to which we will confine ourselves here. It is based on the formation of carbon-14 from nuclear reactions between cosmic rays and nitrogen molecules. The atoms thus formed are immediately incorporated into CO_2 molecules through chemical interaction with ozone.

The last 40 years have been marked by several highlights in terms of developments in instrumentation, advances in accuracy and the diversity of datable samples. Technological advances have clearly enabled radiocarbon to develop to the extent we have seen in recent decades. This was due in part to the development of tandem accelerators in the late 1970s, following Muller's article [MUL 77], which is considered to be the seminal work in the field of direct counting of atoms, rather than beta radioactive decays. The Accelerator Mass Spectrometer (AMS) made it possible to analyze very small quantities of carbon, down to a few milligrams. One of the best-known examples of this was the direct dating of plant carbon used as pigment in ornate caves [VAL 05]. Even today, the two radiocarbon measurement systems, conventional beta decay counting and AMS, coexist. More recently, the miniaturization of mass spectrometers has made it possible to offer compact equipment. MICADAS-type equipment (acronym for *MIni CArbon DAting System*) [SYN 07, BAR 15] can be used in rooms of just a few dozen m².

Libby's hypothesis concerning the constancy of radiocarbon production by cosmic rays was proved wrong as soon as measurement accuracy became sufficiently precise. This was the case from the late 1950s onwards. The reason for this lies in temporal variations in the intensity of cosmic rays and their interactions in the upper atmosphere. Corrections have therefore been made to dating using coupled dendrochronological and radiocarbon analysis of the famous plant cells derived from the growth rings of well-dated trees [FRI 04].

When the GMPCA (Groupe des Méthodes Pluridisciplinaires Contribuant à l'Archéologie or Group of Multidisciplinary Methods Contributing to Archaeology) was set up in 1977, dendrochronology calibration enabled corrections to be made as far back as 10,000 years before present. Thanks to an international effort, the extension of possible comparisons to coral formations, or stalagmitic formations by U/Th and radiocarbon coupling, and the counting of organic varves, now makes it possible to propose corrections over the entire radiocarbon range, i.e. over the last 50,000 years [REI 13].

3.1.2. *Cosmogenic isotopes: the case of carbon-14*

As explained in Chapter 2 (see the end of Application 2.1), nuclear spallation is involved in the interaction of *cosmic ray* particles with matter. Particles with energies ranging from a few MeV to 10^{10} MeV can originate from solar flares and supernovae, while those with energies above 10^{11} MeV are thought to have extragalactic origins [DEG 17].

Cosmic radiation is made up of primary cosmic rays and secondary cosmic rays. *Primary cosmic rays* are made up of 83% protons, 12% light nuclei (mainly helium-4 nuclei), 3% electrons and 1% heavy nuclei, such as carbon, nitrogen and oxygen. The interaction of primary cosmic rays with matter generates cascading reactions that lead to the formation of *secondary cosmic ray* particles. These in turn react with chemical elements in the atmosphere and soil to form new nuclei called *cosmogenic isotopes*. A *cosmogenic radionuclide* is also called a *cosmonuclide*. Among the cosmogenic isotopes used in *radiochronometry* is radiocarbon-14.

The natural production of carbon-14 is formed in the upper layers of the atmosphere [LAL 67, MAS 99]. Cosmic and galactic solar rays contain protons, which react with air molecules to release high-energy neutrons. These neutrons, formed in the upper atmosphere (typically around 70 km altitude), collide with atomic nuclei in the air, such as nitrogen and oxygen. They are slowed down by the numerous neutron–molecule collisions and reach the thermal energy of the gases. They then react with the nuclei of air atoms present in the atmosphere and troposphere to produce carbon-14. Several nuclear reactions are involved in this carbon-14 production system. These include the nuclear reactions ^{14}N (n, p) ^{14}C; ^{16}O (n, 3p) ^{14}C; ^{16}O (n, p, d) ^{14}C and ^{16}O (n, 2p) ^{14}C. Since oxygen is essentially inert to neutrons, nitrogen-14, which is highly quantitative and fairly reactive, is most responsible for the natural production of ^{14}C.

In the upper atmosphere, the impact of a thermal neutron (from secondary cosmic rays) on a nitrogen-14 nucleus produces ^{14}C via the nuclear reaction ^{14}N(n, p)^{14}C, explained as follows:

$$^{14}_{7}N + n \rightarrow {}^{14}_{6}C + p \qquad [3.1]$$

3.1.3. *Radiocarbon-14 in the biosphere*

Formed in the upper layers of the atmosphere, radiocarbon-14 is incorporated into the atmosphere, the ocean and the biosphere. In the biosphere, a living organism assimilates carbon without isotopic distinction. Through oxidation in the air, radiocarbon-14 is found in atmospheric carbon dioxide in the same way as the ^{12}C isotope. Therefore, living organisms absorb atmospheric carbon dioxide from carbon-14 ($^{14}CO_2$), carbon-12 ($^{12}CO_2$) or carbon-13 ($^{13}CO_2$). The proportions of ^{12}C and ^{14}C are assumed to be the same in the atmosphere and in animals and plants, as long as they are alive. When the living organism dies, it ceases to absorb atmospheric $^{14}CO_2$. The proportion of ^{14}C in the dead organism then decays with a half-life of 5,730 years according to the β^- decay mode, without renewing itself.

VOCABULARY CORNER.–

– *Cosmic radiation*: extremely high-energy subatomic particles – mainly protons and atomic nuclei accompanied by electromagnetic emissions – that travel through space and end up bombarding the Earth's surface.

– *Cosmogenic isotopes*: isotopes created by the interaction between a high-energy cosmic ray and an atomic nucleus. This is a spallation reaction caused by cosmic rays. These isotopes are produced in particular in terrestrial materials, such as rocks and soil, in the Earth's atmosphere and in extraterrestrial bodies, such as meteorites (*rocky bodies of extraterrestrial origin that have survived the passage through the atmosphere and are therefore found on the ground*).

3.1.4. *Principle of ^{14}C dating*

Carbon-14 dating is based on the measurement of the $^{14}C/^{12}C$ *isotopic ratio* of the order of 10^{-12} [NDE 11, NDE 12, SÈN 19a, SÈN 19b].

Experimentally, the $^{14}C/^{12}C$ isotope ratio can be determined indirectly by measuring the specific activity due to natural radiocarbon, which is proportional to the $^{14}C/^{12}C$ ratio. However, direct measurement of this ratio by mass spectrometry is preferred, as this technique enables much smaller samples (less than a milligram) to be dated in a minimum of time (less than an hour).

In practice, the carbon extracted from the sample to be dated is first transformed into graphite carbon containing stable ^{12}C (98.93 %) and ^{13}C (1.07%) isotopes, and trace amounts of radioactive ^{14}C. The graphite carbon is then transformed into ions in an ionization chamber and accelerated in the mass spectrometer. This enables the various carbon isotopes to be separated and the $^{14}C/^{12}C$ ratio to be deduced. It should be noted that the ^{14}C *radiochronometer* cannot date samples of living organisms over 50,000 years old, as the $^{14}C/^{12}C$ isotope ratio in these samples is too low to be measured by current techniques. In addition, a certain quantity of sample material, between 5 mg and 2 grams, depending on the material and its state of preservation, is required for the method to be applicable.

To measure ^{14}C radioactivity (the number of electrons emitted per unit of time during counting following β^- decay), there are generally three types of measurement: Gas Proportional Counting (GPC), Liquid Scintillation Counting (LSC) and Accelerating Mass Spectrometry (AMS). GPC and LSC are conventional methods for detecting the radioactivity of the sample under study. Both require at least a few grams of sample and take a few days to complete (approximately 3 days). Figure 3.1 shows the Packard Tri-carb 3170TR/SL Liquid Scintillation Counter, fitted with a $BGO(Bi_4GeO_{12})$ tube for very low background in "Super Low Level" mode, installed at the IFAN Radiocarbon Laboratory of the University Cheikh Anta Diop of Dakar [SÈN 19c].

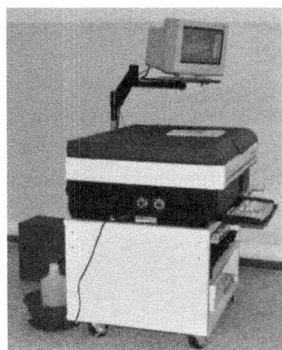

Figure 3.1. *Packard Tri-carb 3170TR/SL liquid scintillation meter*

AMS requires less sample (between 20 and 500 mg). Analysis time is also reduced, usually to a few hours for one sample. This most recent technique enables direct measurement of the number of carbon-14 atoms in the sample in question. In all cases, the sample is destroyed and transformed either into a gaseous carbon compound (CO_2) for GPC, into an organic compound (benzene, C_6H_6) for LSC, or into solid carbon for AMS. Note that for a given organic compound, precise measurements with modern mass spectrometers show that ^{13}C, ^{12}C and ^{14}C isotopes do not behave identically. In other words, during photosynthesis, for example, green plants or tree leaves assimilate atmospheric CO_2 carbon dioxide, preferentially absorbing ^{12}C, then ^{13}C and finally ^{14}C. This shows that *isotopic fractionations* ($^{14}C/^{12}C$ and $^{13}C/^{12}C$) are not the same and can vary from one plant to another.

The carbon-14 dating method was invented in 1945 by American physicist and chemist Willard Frank Libby (1908–1980). In his work, Libby used purified solid carbon distributed in a thin layer, which he placed in a Geiger counter. He then measured the ^{14}C activity of the sample. Libby measured a carbon-14 decay period equal to $5,568 \pm 30$ years [LIB 49]. Today, more precise experiments give a period of 5,730 years. Subsequently, the Dutch scientist Hessel De Vries (1916–1959) made significant improvements to Libby's method by developing a dating technique called proportional gas counting. In this technique, carbon dioxide gas from the sample under study is used for counting [DEV 53].

For the ^{14}C chronometer to be usable, the following conditions must at least be met [SÈN 19c]:

– cosmic ray flux and ^{14}C production must have been constant for at least the last 60,000 years;

– the distribution of ^{14}C in the biosphere must be uniform;

– the absence of chemical or isotopic exchange of the sample's carbon with the carbon present in the vicinity of the sample.

Today, these conditions cannot be met due to anthropogenic pollution. This makes the applicability of ^{14}C dating extremely complex. Similarly, the Bombe effect and the Suess effect can alter the conditions under which the method can be applied. In addition, variations in the Earth's magnetic field play an important role in the production of cosmic rays, which are responsible for the natural production of radiocarbon-14.

Note the effect of lightning on the production of carbon-13 in the ^{13}C/^{12}C isotopic fraction.

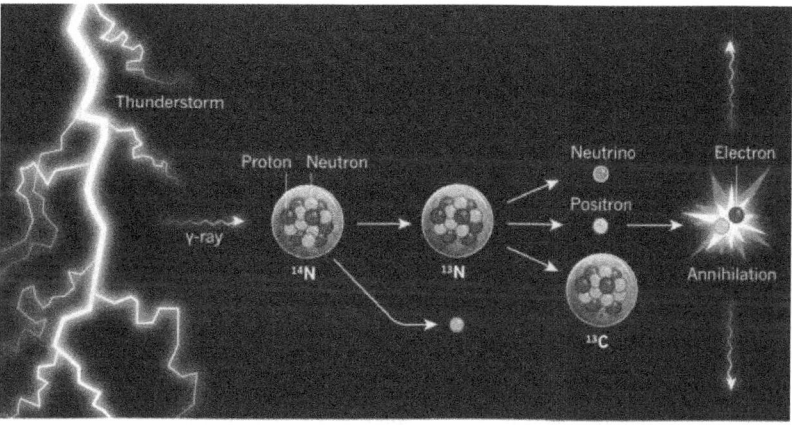

Figure 3.2. *Lightning-induced nuclear reaction with the production of carbon-13 (^{13}C) [DEL 17]*

As shown in Figure 3.2, a gamma ray produced by lightning interacts with a nitrogen-14 (^{14}N) nucleus. This causes the expulsion of a neutron and the formation of an unstable nitrogen-13 (^{13}N) nucleus. The latter decays in the β^+ mode, producing carbon-13 (^{13}C). The positron formed annihilates with a middle electron to produce two gamma photons propagating in directly opposite directions. The reactions involved in the production of ^{13}C can be written as follows:

$$^{14}_{7}N + \gamma \rightarrow {}^{13}_{7}N + {}^{1}_{0}n \qquad\qquad\qquad [3.2a]$$

$$^{13}_{7}N \rightarrow {}^{13}_{6}C + {}^{0}_{1}e + \bar{\nu} \qquad\qquad\qquad [3.2b]$$

$$^{0}_{1}e + {}^{0}_{-1}e \rightarrow \gamma + \gamma \qquad\qquad\qquad [3.2c]$$

Furthermore, obtaining a ^{14}C age requires considerable calibration work. The production rate of this isotope varies over time as a result of the carbon cycle and the modulation of incident cosmic rays by the Earth's magnetic field and solar activity. This ongoing calibration effort is mainly carried out through a combination of approaches, such as dendrochronology, analysis of varved lakes and cross-fertilization of absolute dating methods (^{14}C and ^{234}U/^{230}Th) on speleothems and corals. Steady progress in the calibration of ^{14}C ages [STU 98, REI 04, REI 13] enhances the accuracy and precision of this method and justifies its widespread use. However, this method has a number of limitations. Obtaining a ^{14}C age requires knowledge of the reservoir age, which is not always available. In addition, this method does not allow dating of objects older than 50 ka (1 ka = 1,000 years, see the vocabulary corner below), given the half-life of ^{14}C [SAS 15]. In the literature, we note that radiocarbon ages are also expressed in (kilo) years BP (*Before Present*, before 1950) [FRE 22]. The "kilo year" thus corresponds to the unit ka (kilo annus). Section 3.1.6 presents a complete dossier on the principle of radiocarbon age calibration.

Vocabulary Corner.–

– *Absolute dating*: a dating technique that produces a numerical result, expressed in years. It can concern an event, an object, a geological layer (rock, soil, sediment) or an archaeological level (human remains: tools, bones, pottery, weapons, coins, jewelry, clothing, etc.).

– *Relative dating*: this encompasses all dating methods used to chronologically order geological or biological events in relation to one another. It complements or opposes absolute dating.

– The conventional *carbon-14 age*: age of a sample of organic matter, expressed in "Before Present" (BP) years. Libby set 1950 as the reference date from which to measure the time elapsed since the death of the organism. Thus, by convention, the present corresponds to the year 1950 (year zero of the radiocarbon calendar). A 10,000-year-old BP fossil would be 10,000 years older than 1950.

– *Unit ka*: the notion of time in Earth Sciences is essential. Nevertheless, there is often confusion between absolute time and duration in the literature. To bring Earth Sciences into line with the SI (*International System of Units*), Holden et al [HOL 11] proposed in 2011 the definition of the annus as a unit of time, with the symbol "a". The *annus* corresponds to 3.1556925445×10^7 seconds and is therefore defined as a fixed multiple of the second, which is the SI unit of time. To be consistent with SI unit usage, an age and a duration must be expressed in the same unit. The symbol "a" can be completed with the prefixes k ($\times 10^3$), M ($\times 10^6$) and G ($\times 10^9$). There is no space between the prefix and the unit, i.e. ka, Ma and Ga to designate thousand, million and billion years respectively. The unit BP is reserved for radiocarbon dates, which must be preceded by "cal" (e.g. cal. BP) when the date is calibrated [NOM 17]. Note that "annus" is a Latin word meaning circle, year or age.

– *Dendrochronology*: a method that makes it possible, in particular, to date pieces of wood to the nearest year by counting and analyzing the morphology of tree rings (growth rings).

– *Varved lakes*: a varve is a sedimentary layer or stratum deposited over the course of a year at the bottom of a lake. This stratification is due to annual variations in climate (seasons). The study of varves is useful for dating recent climatic and geological events. A *sediment* is a collection of particles suspended in water, atmosphere or ice, which has been deposited by gravity, often in successive layers or strata.

– *Speleothems*: more commonly known as concretions, these are mineral deposits precipitated in a natural underground cavity (an example of speleothems is stalagmites, made up of limestone dissolved in drops of water falling on the ground).

– *Corals*: marine animals, often living in colonies, with an orifice surrounded by tentacles enabling them to capture prey for feeding. For more details, see the note at the end of section 3.4.3 dedicated to coral dating.

3.1.5. *Age correction, radiocarbon age and calendar age*

Since the 1960s, the rate of natural radiocarbon production has varied over time, thus affecting isotopic fractionations ($^{14}C/^{12}C$ and $^{13}C/^{12}C$). In the magnetosphere, variations in the Earth's magnetic field exert a shielding effect on the primary cosmic radiation. This causes fluctuations in the rate of thermal neutron production, and thus in the amount of cosmogenic carbon-14 that depends on it. In addition, effects such as the *"Bomb" effect* and the *"Suess" effect* help to modify the concentration of radiocarbon in the atmosphere. The "Bomb" effect is linked to the increase in ^{14}C emissions (increase in the $^{14}C/^{12}C$ ratio) due to nuclear weapons testing. The "Suess" effect refers to the combustion of fossil fuels (coal, oil, natural gas, etc.) in the atmosphere, producing fossil CO_2. This altered the isotopic composition of fossil carbon, leading to a depletion of the ^{14}C level (lower $^{14}C/^{12}C$ ratio) in industrial zones [SÈN 19c]. Finally, we might mention the *reservoir effect*, reflecting the fact that oceanic and atmospheric concentrations of radioactive ^{14}C are not homogeneous.

According to the above facts, the global quantity of carbon-14 in the biosphere is not constant over time. To correct Libby's model, calibration curves are constructed by comparing carbon-14 dates with those obtained by other methods, such as dendrochronology. Using these curves, *BP ages* are transformed into *calibrated* or *calendar ages* expressed as chronological intervals associated with a probability percentage [STU 98, JUL 03, FON 04]. The idea behind calibrated ^{14}C ages is to reconstruct the variation of the $^{14}C/^{12}C$ ratio in the past, in order to correct ^{14}C ages. Knowing the initial ^{14}C content of a sample means studying the evolution of the $^{14}C/^{12}C$ ratio in the atmosphere. Comparing a ^{14}C age (*radiocarbon age*) with a real age (*calendar age*) of the same fossil makes it possible to reconstruct the $^{14}C/^{12}C$ ratio of the period [FON 04]. Thus, the comparison between calendar age (T_{true}) and radiocarbon age (T_{14C}) gives direct access to the atmospheric isotopic ratio ($^{14}C/^{12}C)_{sample}$ of the epoch considered, according to the formula [SÈN 19c]:

$$T_{true} - T_{14C} = \frac{1}{\lambda} \ln \left(\frac{(^{14}C/^{12}C)_{sample}}{(^{14}C/^{12}C)_{1950}} \right) \quad [3.3]$$

If we disregard variations in the $^{14}C/^{12}C$ ratio and consider this ratio to be constant, we can estimate the age of a fossil sample of living matter using the exponential decay law $A(t) = A_0 e^{-\lambda t}$. In this law, A_0 denotes the activity of the sample taken from a prehistoric site and $A(t)$ denotes the activity of an identical sample of the same mass taken at a recent date t (see Application 3.1).

APPLICATION 3.1.–

The activity of a piece of charcoal found in a prehistoric cave is equal to 0.03 Bq. A recent charcoal sample of the same mass has an activity of 0.2 Bq. From these experimental measurements, deduce the approximate age of the piece of charcoal.

Information: $2^{2.737} = 6.667$. For ^{14}C: $T = 5,730$ years.

ANSWER.–

Let t_a be the age of the piece of charcoal. At $t_0 = 0$, the activity $A_0 = 0.2$ Bq and $t = t_a$; $A(t_a) = 0.03$ Bq. Knowing that the decay constant $\lambda = \ln2/T$, we obtain:

$$A(t_a) = A_0\, e^{-\lambda t_a} = A_0\, e^{-\ln 2\, t_a / T} \qquad\qquad [3.4a]$$

This results in:

$$t_a = \frac{\ln\left[A_0 / A(t_a) \right]}{\ln 2} T \qquad\qquad [3.4b]$$

NA: $t_a = \ln\,[0.3/0.03] \times 5{,}730/\ln 3 = 15{,}683$ years.

NOTE.–

Cheikh Anta Diop (1923–1986) [DIO 74], a Senegalese historian, anthropologist, Egyptologist and politician, defended his doctoral thesis in anthropology in 1966. He went on to specialize in nuclear physics at the Collège de France's nuclear chemistry laboratory. In August 1936, the French Institute of Black Africa was created by Order No. 1945/E of the Governor General of French West Africa, Jules Brévié (1880–1964). In 1957, the University of Dakar was founded, the oldest French-speaking university in sub-Saharan Africa. After the independence of African countries, the French Institute of Black Africa was integrated into the University of Dakar in 1963 and changed its name to become the "Institut Fondamental d'Afrique Noire" (IFAN) of the University of Dakar. In 1966, in collaboration with the Radiocarbon Laboratory of the French Atomic Energy Commission in Gif-sur-Yvette, Diop created the first African laboratory for the dating of archaeological fossils with carbon-14 at IFAN. In March 1987, the University of Dakar was renamed Cheikh Anta Diop University.

3.1.6. *Calibrating radiocarbon ages: reasons for calibration? How to calibrate?*

This is a lesson designed to teach readers working in the field how to calibrate radiocarbon ages with R [FRE 22]. It covers a step-by-step explanation of how to calibrate a set of dates and how to explore and present the results. The principle of radiocarbon was explained in section 3.1.2. We therefore omit it in this lesson.

– Why calibrate radiocarbon ages?

In the second half of the 20th century, as progressively older objects were dated, the gap between measured and expected ages became increasingly apparent.

Contrary to Libby's postulate, ^{14}C content in the atmosphere is not constant over time, which partly explains the observed variations. Atmospheric ^{14}C content varies as a function of natural phenomena (variations in the Earth's magnetic field, solar activity, volcanic activity, the carbon cycle, etc.) and human activity. These phenomena can be contradictory: the use of fossil fuels releases very old carbon and tends to reduce the relative ^{14}C content (Suess effect); conversely, atmospheric nuclear testing has produced large quantities of ^{14}C.

The chronometer constituted by the radiocarbon method therefore does not have a regular rhythm (as atmospheric ^{14}C content varies over time). As a result, radiocarbon ages (hereafter referred to as *conventional ages*) belong to a time frame of their own.

Nevertheless, using Libby's postulate remains the only accessible way to estimate the initial amount of ^{14}C at system closure. It is therefore necessary to perform a *calibration* operation to transform a conventional age into a calendar age. This operation is performed using a curve whose estimation is regularly updated by the scientific community. The calibration curve is constructed by dating samples both by radiocarbon and by an independent method, thus providing a table of equivalence between radiocarbon time and calendar time (Figure 3.3b).

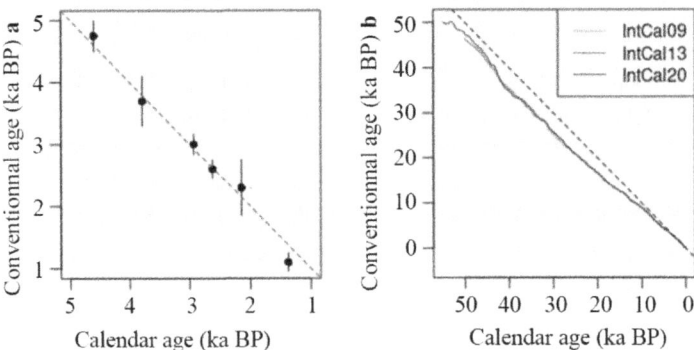

Figure 3.3. *Radiocarbon ages versus expected calendar ages. (a) Curve of radiocarbon ages of archaeological objects whose calendar age is known by independent methods [ARN 49]. The 1:1 line, for which a conventional age is equal to a calendar age, is shown as dashed lines. (b) IntCal09, IntCal13 and IntCal20 calibration curves [REI 09; 13, 20]. The deviation from the 1:1 line (dashed line) is more pronounced for older ages*

– How to calibrate

Thanks to the equivalence table between radiocarbon time and calendar time, the calibration process is relatively straightforward. In reality, however, the calibration process becomes more complex when we take into account the errors inevitably associated with physical measurements.

A conventional age (noted here as *t*) is the result of a measurement and, as there is no such thing as a perfect measurement, it is always accompanied by a term corresponding to the analytical uncertainty (Δt) and expressed in the form $t \pm \Delta t$ (the age, *plus or minus* its uncertainty). This uncertainty results from the combination of different sources of error within the laboratory: it is a *random* uncertainty inherent in measurement.

A conventional age is thus an estimator of the true radiocarbon age of the dated object. If the same sample is dated a large number of times, its value is likely to vary, and there is very little chance that it will coincide exactly with the true radiocarbon age. It is therefore preferable to focus on an interval that is highly likely to contain the true (unknown) value of the conventional age. In fact, uncertainty characterizes the dispersion of values that could reasonably be attributed to the true age. A conventional age is the result of a random process, radioactive decay, and can be modeled using a specific probability distribution: the normal distribution.

Only two parameters are needed to characterize the distribution of values according to a normal distribution: the mean μ (central tendency) and the standard deviation σ (dispersion of values). The properties of the normal distribution are such that the interval defined by $\mu \pm \sigma$ contains 67% of the values. If we multiply the standard deviation by two, the interval $\mu \pm 2\sigma$ contains 95% of the values (Figure 3.4).

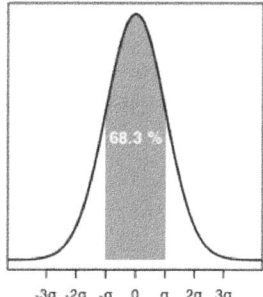 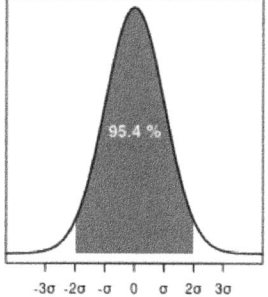 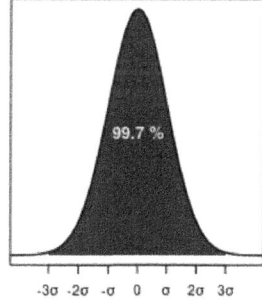

Figure 3.4. *Normal distribution with mean 0 and standard deviation 1, with normality ranges at the 68%, 95% and 99% confidence levels. The distribution of values is such that dispersion is symmetrical around the central tendency*

Therefore, if we express the uncertainty of a conventional age as a function of the standard deviation, there is a 68% chance that the 1σ interval will contain the true conventional age. Similarly, the 2σ interval has a 95% chance of containing the true conventional age. The interval at 1σ is less dispersed, but less likely to be accurate than at 2σ: the range of values retained is tighter, but less likely to contain the true conventional age.

The most basic approach to calibrating a radiocarbon age is to intercept the calibration curve between the uncertainty bounds ($t - \Delta t$ and $t + \Delta t$ in the 1σ case) to obtain the corresponding calendar age interval. This is illustrated in Figure 3.5, which shows the calibration of a conventional age by intercepting a calibration curve (solid train), whose uncertainty is represented by a gray band. Conventional and calendar ages are shown at 1σ (black bands) and 2σ (hatched bands).

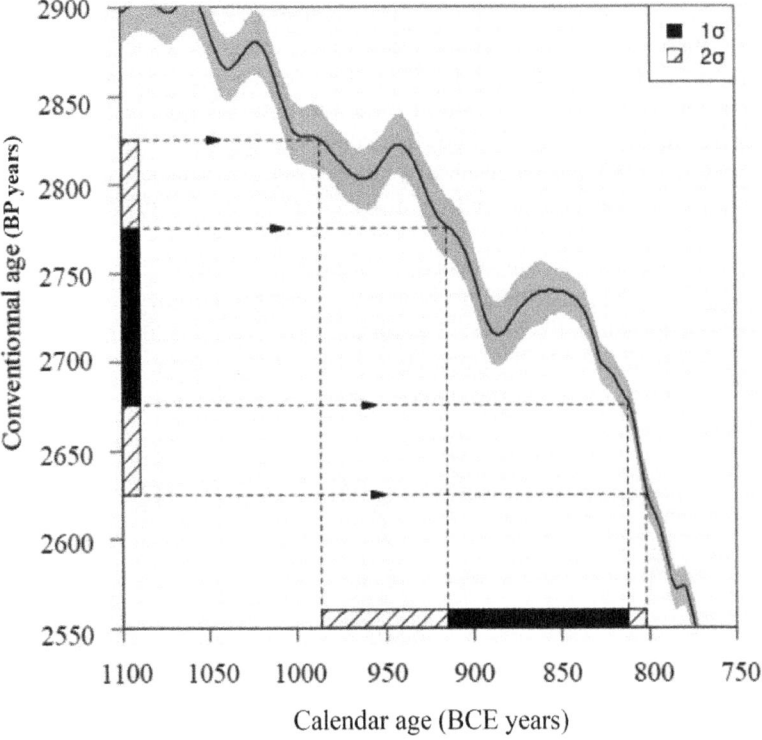

Figure 3.5. *Calibration of a conventional age of 2,725 ± 50 years BP by intercepting the IntCal20 calibration curve. BP: before present, BCE: before the Common Era*

However, this approach does not take into account the fact that a radiocarbon age is described by a normal distribution. Within the range defined by the radiocarbon age plus or minus its uncertainty, not all values have the same probability of coinciding with the true radiocarbon age, and calibration by simple interception assumes the opposite. In fact, the current approach is to also take into account the normal distribution of radiocarbon ages. This is sometimes referred to as *probabilistic calibration*. This calibration method uses numerical methods, and the resulting distribution of calendar ages is not equiprobable (Figure 3.6).

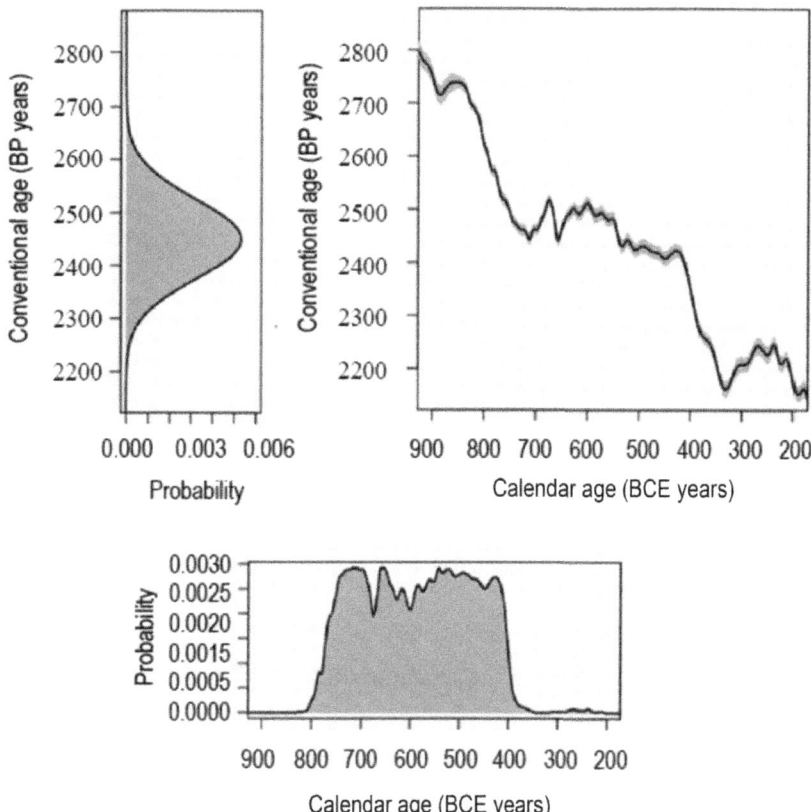

Figure 3.6. *Distributions of a radiocarbon age of 2,450 ± 75 years BP before and after calibration, top left and bottom right respectively. Top right: extract from the IntCal20 calibration curve (solid line) and associated error (grey band)*

The interval to which a calendar age belongs results from the uncertainty of the conventional age, the shape of the calibration curve and the uncertainty associated with the latter. This interval, within which the calendar age has a given probability of being included, can be obtained in two distinct ways (Figure 3.7):

– *Highest posterior density interval (HPDI)*: the bounds of the interval correspond to the regions of the distribution whose cumulative probability is greater than a given threshold.

– *Credibility interval*: the bounds of the interval correspond to the distribution quantities.

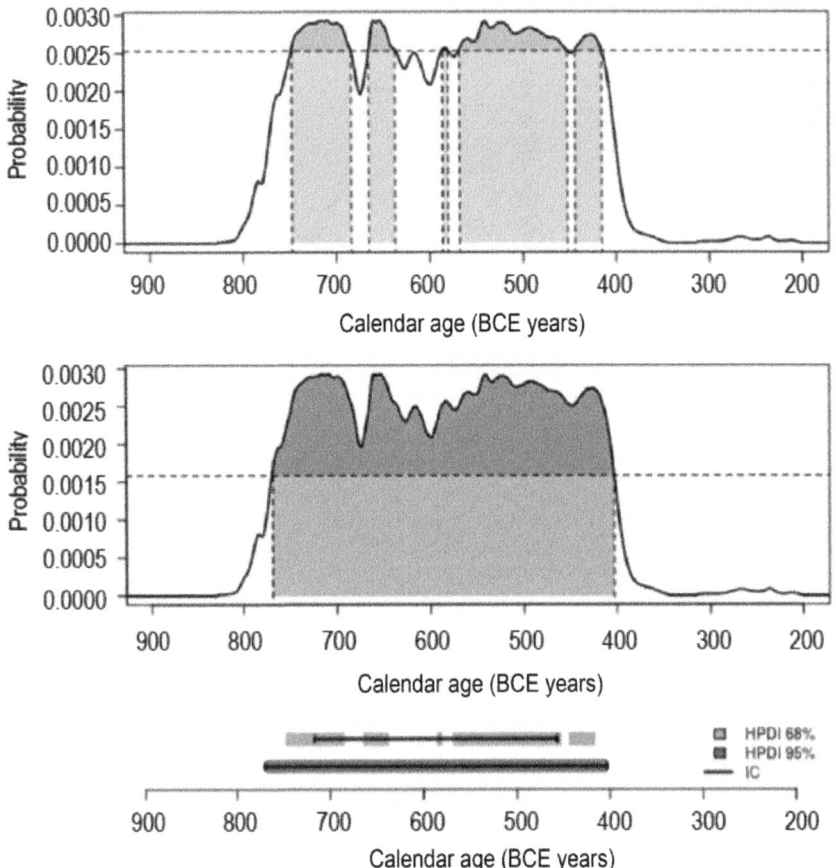

Figure 3.7. *Estimation of calibrated intervals. The top two graphs illustrate the estimation of the highest density regions at 68% and 95%. The lower graph compares the HPDIs thus obtained with the corresponding credibility intervals (solid lines)*

If the distribution of a calibrated age is multimodal (sawtooth), the highest density interval often corresponds to the union of several disjoint intervals. This is in contrast to the credibility interval, which always provides a continuous range of values. The higher-density interval is therefore often more informative, which is why it is commonly used to present calibrated results.

Certain periods are more or less suitable for radiocarbon dating, depending on the shape of the curve. The least favorable case is the existence of plateaus within the calibration curve. A typical case is the Iron Age plateau (Figure 5.4). For example, a conventional age of 2,450 ± 75 years BP corresponds, once calibrated at 95% (HPDI), to a calendar age between 2,719 and 2,353 years BP (i.e. 769–403 BCE).

Therefore, despite a conventional age with a fairly low uncertainty (3%), the corresponding calendar age has a 95% chance of lying within a time interval that covers almost the entire early Iron Age (Figure 3.6). If we calibrate at 68% (HPDI), we are faced with another difficulty linked to oscillations in the calibration curve. The calendar age has a 68% chance of belonging to the union of intervals 748–684 (18%), 665–637 (8%), 586–580 (2%), 568–452 (32%) and 444–415 (8%) before our era, and not to a unique interval (Figure 3.7).

In some contexts, it is common to keep calibrated ages expressed in BP years. In such cases, it is advisable to specify *cal* BP to avoid any confusion on the part of the reader. These calendar ages in BP years can be converted to dates before or after our era (BCE/CE, *before the Common Era/Common Era*). To do this, simply use the following calculation rule.

To convert a *calibrated age* (noted x) expressed in BP years into BCE/CE years, knowing that there is no year 0 in the Gregorian calendar:

– If the calibrated age is less than 1950 BP: 1950–xCE.

– If the calibrated age is greater than or equal to 1950 BP: 1949–xBCE.

If misunderstood, these peculiarities can quickly lead to over-interpretation. During the study of a dating corpus, or when it is published, it is therefore particularly important to present all the data and choices involved in obtaining the calendar ages. The use of open-source tools promotes both transparency and reproducibility of results, two particularly important aspects when it comes to calibrating radiocarbon ages.

Numerous tools are now available for calibrating O radiocarbon ages. The R language offers an interesting alternative. Distributed under a free license, it favors reproducibility and enables radiocarbon age processing to be integrated into broader studies (spatial analysis, etc.). For a concrete application of radiocarbon age calibration with R, the interested reader is invited to consult reference [FRE 22].

3.2. Potassium–argon (K–Ar) dating

The dating method based on potassium-40 filiation is used in geochronology to date the formation of rocks containing potassium, obviously, and capable of retaining the Argon-40 formed. One of the first mentions of the possibility of using this filiation in geochronology was published in 1948 by physicists Aldrich and Nier of the University of Minnesota [GIL 82, ALD 48]. Today, it is the method of choice for volcanic chronology [LEF 94, MCD 14]. The method is based on the measurement of the ^{40}K/^{40}Ar ratio in the mineral under study, a ratio that changes and decreases over time. Two techniques are used: Potassium–Argon and Argon–Argon.

In the case of Potassium–Argon, the measurement of argon-40 produced over time is carried out on an aliquot of the sample by mass spectrometry. The measurement of potassium is carried out on a second aliquot by atomic emission spectrometry after the minerals have been put into solution [GUI 18].

3.2.1. *Principle of dating*

Potassium decay was first demonstrated in 1948 by Aldrich and Nier [ALD 48], who set out to measure the ^{40}Ar/^{36}Ar ratio of various potassium minerals. Potassium-40, with radioactive half-life $T = 1,248 \times 10^9$ years, decays at [SAK 22]:

– 88.8% to the fundamental calcium-40 level;

– 11.2% to the argon-40 fundamental level;

– 0.001% to the excited level (at 1460.8 keV) of argon-40.

The decay energy diagram for potassium-40 is shown in Figure 3.3 [SAK 22].

The K–Ar dating method exploits the decay property of the father nucleus ^{40}K to produce the *radiogenic* son nucleus ^{40}Ar. This method covers the entire time scale from the origin of the universe to 30,000 years BP [DIO 74]. From a chronological point of view, the exponential decay of potassium-40 was demonstrated in 1948.

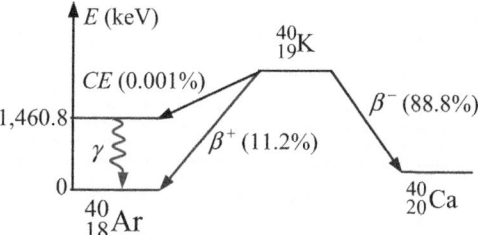

Figure 3.8. *Potassium-40 decay energy diagram.*
Here are its three disintegration paths

In general, the K–Ar method is based on the determination of the ^{40}Ar/^{40}K *isotope ratio* in a given mass of rock sample. This method is ideally suited to the dating of volcanic rocks. Potassium is one of the eight most abundant elements in the Earth's crust (Table 3.1).

Element	O	Si	Al	Fe	Ca	Na	Mg	K
%	46.6	27.7	8.1	5.0	3.6	2.8	2.3	2.1

Table 3.1. *The eight elements that make up more than 98% of the composition of the rocks in the Earth's crust*

In magma containing ^{40}K, some or all of the ^{40}Ar formed can escape through outgassing. During a volcanic eruption, the magma reaches the surface, then cools and solidifies. The ^{40}Ar is then trapped in the magmatic rock, where it can no longer escape. This accumulates in the lava crystallized by the disintegration of ^{40}K. The K–Ar age of a lava thus corresponds to its emplacement age, i.e. the age of the eruption [SAS 15]. Determination of this age by the K–Ar *radiochronometer* is based on the *age equation* established in section 3.2.3.

3.2.2. Basic assumptions for the K–Ar radiochronometer

The use of the K–Ar clock is based on several assumptions. Knowledge of these assumptions is fundamental, as they play a major role at different levels of sample analysis. This enables us to take a more pertinent, critical look at the results obtained. Let us consider the following basic assumptions [SAS 15]:

– At the moment of its formation (i.e. $t = 0$), the sample is considered to be devoid of any ^{40}Ar*. This means that only ^{40}Ar of atmospheric origin is present in the system when it closes, i.e. a $^{40}Ar/^{36}Ar$ ratio equal to that of air (= 295.5) at $t = 0$. Otherwise, the ages obtained would be overestimated due to an excess of ^{40}Ar. These excesses of ^{40}Ar are identifiable by the $^{40}Ar/^{39}Ar$ method, but not by the $^{40}K–^{40}Ar$ method, for which they constitute a real limit.

– The sample's formation time must be negligible compared with its age. In the case of volcanic rocks, which are the subject of this study, this assumption is validated by the fact that their cooling time is negligible, compared to the age range of the samples studied (20–200 ka).

– The sample studied must have evolved in a closed system since its formation and must not have undergone any alteration. In other words, the K and Ar content of the sample must not have been affected by any event since its formation. This assumption can be verified by the $^{40}Ar/^{39}Ar$ method, but not by the K–Ar method. In order to best constrain this hypothesis, only the least altered samples are taken in the field. The absence or degree of alteration is verified in the laboratory by macroscopic and microscopic observations.

– ^{40}K decays at a constant rate which does not depend on the physical conditions of the system. The decay constants for ^{40}K are those given in Table 3.3. The $^{40}K/K_{total}$ ratio is constant in natural materials. This is because ^{40}K is not measured directly, but via K_{total} (as $^{40}K–^{40}Ar$) or $^{39}Ar_k$ ($^{40}Ar/^{39}Ar$), combined with isotopic abundances.

Note that in the text above, the notation Ar* stands for radiogenic argon (a radiogenic isotope is one produced by a decay reaction). Table 3.2 shows the abundance (molar percentage) of argon and potassium isotopes [NIE 50, GAR 75, SAS 15].

Isotope	^{39}K	^{40}K	^{41}K	^{40}Ar	^{38}Ar	^{36}Ar
%	93.2581	0.01167	6.7302	99.600	0.0632	0.3364

Table 3.2. *Abundance of argon (Ar) and potassium (K) isotopes*

3.2.3. *Age equation*

Decay of ^{40}K produces ^{40}Ar (also known as $^{40}Ar*$) and ^{40}Ca. N_{0K} is the number of potassium-40 nuclei present in the lava immediately after crystallization. The number $N_K(t)$ of ^{40}K nuclei remaining in the crystallized lava at time t is:

$$N_K(t) = N_{0K}e^{-\lambda t} \Rightarrow N_{0K}(t) = N_K(t)e^{\lambda t} \qquad [3.5]$$

Let $N_{Ar}(t)$ and $N_{Ca}(t)$ be respectively the numbers of ^{40}Ar and ^{40}Ca nuclei formed in the crystallized lava. Taking advantage of the conservation of the number of nuclei, we obtain:

$$N_{0K} = N_K(t) + N_{Ar}(t) + N_{Ca}(t) \qquad\qquad [3.6]$$

Using [3.5], we find according to [3.6]:

$$N_{Ar}(t) + N_{Ca}(t) = N_{0K}-N_K(t) = N_K(t)\,[e^{\lambda t}-1] \qquad [3.7]$$

Let us say:

$-\lambda_{Ar}$: the decay constant of ^{40}K which produces ^{40}Ar;

$-\lambda_{Ca}$: the decay constant of ^{40}K which generates ^{40}Ca;

$-\lambda = \lambda_{Ar} + \lambda_{Ca}$: the total decay constant of ^{40}K.

The values of the above decay constants are shown in Table 3.3 [STE 77, SAS 15]:

λ_{Ar}	$(5.808 \pm 0.004) \times 10^{-11}\ y^{-1}$
λ_{Ca}	$(4.963 \pm 0.009) \times 10^{-10}\ y^{-1}$
λ	$(5.543 \pm 0.010) \times 10^{-10}\ y^{-1}$

Table 3.3. ^{40}K decay constants

Let $N_{KAr}(t)$ be the number of potassium-40 nuclei likely to decay to argon-40 and $N_{Ar}(t)$ the number of accumulated argon-40 nuclei. This gives:

$$N_{Ar}(t) = N_{0K}-N_{KAr}(t). \qquad\qquad [3.8a]$$

with:

$$N_{KAr}(t) = N_{0K}e^{-\lambda_{Ar}t}. \qquad\qquad [3.8b]$$

Using [3.8b], equation [3.8a] yields:

$$N_{KAr}(t) = N_{0K}(1 - e^{-\lambda_{Ar}t}). \qquad\qquad [3.8c]$$

Knowing that $\lambda_{Ar} \ll 1$ (see Table 3.2), we obtain by expanding [3.8c] to the first order:

$$N_{Ar}(t) \approx N_{0K} \lambda_{Ar} t. \tag{3.8d}$$

Similarly, for calcium-40, we find ($\lambda_{Ca} \ll 1$):

$$N_{Ca}(t) = N_{0K}(1 - e^{-\lambda_{Ca} t}) \approx N_{0K} \lambda_{Ca} t. \tag{3.8e}$$

Using [3.8d] and [3.8e], we obtain approximately:

$$\frac{N_{Ca}(t)}{N_{Ar}(t)} = \frac{\lambda_{Ca}}{\lambda_{Ar}} \tag{3.9}$$

From [3.7]:

$$N_{Ar}(t)\left[1 + \frac{N_{Ca}(t)}{N_{Ar}(t)}\right] = N_K(t)\left[e^{\lambda t} - 1\right] \tag{3.10}$$

Considering [3.9], the age equation for the K–Ar radiochronometer is written as:

$$N_{Ar}(t)\left[1 + \frac{\lambda_{Ca}}{\lambda_{Ar}}\right] = N_K(t)\left[e^{\lambda t} - 1\right] \tag{3.11}$$

Knowing that $\lambda = \lambda_{Ar} + \lambda_{Ca}$, we obtain:

$$\frac{\lambda}{\lambda_{Ar}} = 1 + \frac{\lambda_{Ca}}{\lambda_{Ar}} \tag{3.12}$$

Using [3.11] and [3.12], we derive the isotopic ratio $^{40}Ar/^{40}K$. Let:

$$\frac{N_{Ar}(t)}{N_K(t)} = \frac{\lambda_{Ar}}{\lambda}\left[e^{\lambda t} - 1\right] \tag{3.13}$$

If T denotes the half-life of ^{40}K, the age t_a of the volcanic eruption is given by the definitive relationship:

$$t_a = \ln\left\{\left(\frac{N_{Ar}(t_a)}{N_K(t_a)} \times \frac{\lambda}{\lambda_{Ar}}\right) + 1\right\} \times \frac{T}{\ln 2} \tag{3.14}$$

Recall that in relation [3.14], Ar = Ar*, i.e. ^{40}Ar from ^{40}K decay (the asterisk "*" is used to indicate that this is radiogenic Ar).

APPLICATION 3.2.–

The period of potassium-40 is $T = 1.3 \times 10^9$ years. During a volcanic eruption, lava in contact with air loses argon-40 (degassing); at the time of eruption, the lava no longer contains argon. Analysis of a 1 kg basalt sample shows that it contains 1.4900 mg potassium-40 and 0.021 8 mg argon-40 [SAK 16a]. Estimate the date of the volcanic eruption.

ANSWER.–

From Table 3.2, we derive the decay constant for argon-40: $\lambda_{Ar} = 5.808 \times 10^{-11}$ year^{-1}. The decay constant for potassium-40 is $\lambda_K = 5.332 \times 10^{-10}$ year^{-1}. For isobars (same mass number A), the ratio of their numbers of nuclei is equal to the ratio of their masses (remember that: $m/A = N/N_A$). Using [3.14], we find:

$$t_a = \ln\left\{\left(\frac{0.0218}{1.4900} \times \frac{53.32}{5.808}\right) + 1\right\} \times \frac{1.3 \times 10^9}{\ln 2} = 2.4 \times 10^8 \text{ years.}$$

3.2.4. Atmospheric correction

Relation [3.14] is obtained without any *atmospheric correction*, since Ar from the atmosphere is not taken into account. In order to properly exploit the results obtained with the K–Ar radiochronometer, atmospheric correction is necessary.

The total ^{40}Ar contained in a lava has two origins: radiogenic and atmospheric. Therefore [SAS 15]:

$$^{40}\text{Ar}_{total} = {}^{40}\text{Ar}_{at} + {}^{40}\text{Ar}^*$$ [3.15]

In this relationship (expressed in terms of argon-40 concentration: ^{40}Ar = [^{40}Ar] in mol · g^{-1}):

– ^{40}Ar$_{total}$: ^{40}Ar total sample content;

– ^{40}Ar$_{at}$: ^{40}Ar from the atmosphere;

– ^{40}Ar*: radiogenic ^{40}Ar from the decay of ^{40}K.

Experimentally, atmospheric correction is achieved by introducing an aliquot of reference atmospheric argon through an air bomb connected to the mass spectrometer. This dose of air is measured under the same pressure conditions as the sample to be dated. Comparison of the ratios (^{40}Ar/^{36}Ar) of the sample and the air dose, i.e. (^{40}Ar/^{36}Ar)$_{atm}$, gives the ^{40}Ar* value of the sample (in equation [3.15], Ar corresponds to Ar*). Taking atmospheric correction into account, we obtain [SAS 15]:

$$^{40}Ar* = \frac{\dfrac{^{40}Ar_{sample}}{^{36}Ar_{sample}} - \dfrac{^{40}Ar_{atm}}{^{36}Ar_{atm}}}{\dfrac{^{40}Ar_{sample}}{^{36}Ar_{sample}}} . \qquad \text{[3. 16.a]}$$

VOCABULARY CORNER.–

– *Radiogenic isotope*: nuclide resulting from nuclear decay reactions.

– *Aliquot*: proper divisor for a natural number. Example 5 is an aliquot of 25.

3.2.5. *Preparing samples for K–Ar dating*

Before any laboratory analysis, lava samples must be carefully prepared. No measurement can give an acceptable result if the sample analyzed is not of good quality. These pre-measurement steps are therefore essential to obtain a robust age for the lava samples studied. Sample selection and preparation is the same for both the K–Ar and ^{40}Ar/^{39}Ar methods. Sample preparation follows the measurement protocol detailed in Guillou et al. [GUI 98]. This follows a three-step protocol: grinding-sieving, cleaning and mineralogical sorting [SAC 15].

– Grinding–Sieving

After removing the outer parts of the sampled block (potentially the most damaged), it is cut into cubes of around 4–5 cm on each side using a circular saw with diamond chips on the blade. The grinding stage is carried out by decreasing the gap between the crusher jaws, starting at 15 mm and ending at 0 mm, passing through 11 mm, 8 mm and 5 mm.

After each grinding stage, the shredded material is passed through a sieve column comprising two sieves, the first with a 250 μm mesh and the second with a 125 μm mesh. The fraction between 125 μm and 250 μm is retained. The fraction greater than 250 μm passes back through the grinder with a reduced gap between the jaws, and then the grinded material is sieved again. The resulting sample is then of a suitable particle size (between 125 and 250 μm) for analysis.

– Cleaning

After sieving, the sample is rinsed under distilled water to remove most of the dust still adhering to the grains. The sample is then immersed in a beaker containing 99.8% acetic acid. The beaker is then placed in an ultrasonic bath at 60°C for 45 minutes. The purpose of this step is to dissolve any weathering phases present. The sample is then thoroughly rinsed with distilled water and cleaned with alcohol and acetone.

– Mineralogical sorting

Lava does not consist solely of pure mesostasis. It may also contain numerous *phenocrysts*, formed long before the lava cooled, as well as *xenocrysts*. These crystals are potential carriers of excess argon, particularly *olivines* and *pyroxenes* [LAU 94], and may therefore distort the age of the sample if not removed. To this end, different types of sorting are successively carried out, as follows.

– Magnetic sorting

The sample is placed under a magnet. The mesostase, rich in magnetic minerals, is attracted by the magnet. The xenocrystals, on the contrary, remain in place. Once this initial separation is complete, the magnetic phase is recovered for the next sorting stage.

– Densitometric sorting

This step refines the separation undertaken during magnetic sorting. Indeed, some phenocrysts may have been dragged along with the mesostase. For this sorting step, the sample powder is introduced into a funnel containing diiodomethane ($d = 3.32$ g·cm^{-3}). With the gradual addition of acetone ($d = 0.791$ g·cm^{-3}), the density of the liquid decreases. The crystals, denser than the mesostasis, sink to the bottom of the funnel. The mesostasis continues to float. At ad-hoc density (i.e. 2.9–3.0 for basalts), the mesostase is recovered and rinsed with acetone, then oven-dried. The sample is then ready for analysis. A binocular loupe is used to check the purity of the sample obtained.

– Pitting

Should binocular inspection reveal the presence of certain *phenocrysts*, grain-by-grain pitting can be carried out under binocular loupe, prior to analysis. A significant quantity of non-magnetic minerals, potential carriers of excess argon, can be observed. This confirms the importance of these various sortings.

Vocabulary corner.–

– *Mesostasis*: in volcanic rocks, the term *mesostasis* refers to the vitreous or very finely crystalline material filling the spaces between the crystals.

– *Olivine*: a mixed crystal formed from a magnesium (Mg_2SiO_4) and iron (Fe_2SiO_4) silicate mineral, in which the magnesium member is generally dominant.

– *Phenocrystal*: a large crystal in a rock, often magmatic, visible to the eye.

– *Pyroxene*: a ferromagnesian mineral (presence of Fe and Mg in their chemical composition), whose color varies according to its chemical composition. It is the essential crystal of magmatic rocks and of many metamorphic rocks (*rocks that have undergone mineralogical and structural transformation as a result of increased temperature and pressure*).

– *Xenocrystal*: a crystal included in a magmatic rock but not originating from the magma that formed the rock (e.g. quartz crystals included in a silica-deficient lava).

– *Xenolith*: it is a fragment of rock that is surrounded by another rock (foreign rock). It occurs mainly in *igneous rocks* when the magma is fluid enough to flow around more solid rocks.

– *Lithic fragment*: fragments of older rocks embedded in a more recent rock, usually sedimentary but sometimes from a volcanic context.

– *Igneous rocks*: these rocks result from the cooling and crystallization of magma. Igneous rocks are the most common rocks in the Earth's crust. The word "igneous" comes from the Latin igneus, meaning "from fire". Magma is a viscous liquid formed from molten rock and contains gases dissolved at very high temperatures. It is the source of igneous rocks. Magma is carried to the Earth's surface, where it cools and solidifies. Crystals form in a disorderly fashion, with no particular orientation. Rocks resulting from this process are called igneous.

3.2.6. *Experimental protocols for potassium and argon measurements*

In the case of the K–Ar radiochronometer dating method, potassium and argon are measured separately using conventional measurement techniques. Potassium is measured chemically as a solid element, while argon is measured by mass spectrometry as a gas. The principle of determination of these two elements is described below [CAS 82, GIL 86, GIL 06, GOU 75]. For a full description of the experimental protocols, the reader is invited to consult the references given.

– Measuring potassium content

Potassium is generally measured chemically, as it is one of the major constituents of rocks. In fact, radioactive potassium-40 currently represents 0.01167% of natural potassium (see Table 3.2). Determining the amount of natural potassium in the rock or mineral to be dated therefore enables us to deduce its current potassium-40 content per gram of sample. This is the N_K value that we will enter into the age equation. This shows that the analytical uncertainty of this potassium assay is directly transferred to the calculated age value. The potassium assay must therefore be carried out with the utmost precision, particularly for older samples (> 1 My), when the quantity of argon accumulated is such that the assay becomes increasingly precise.

The most commonly applied measurement technique is flame spectrometry. Potassium, an easily excitable alkaline element, is easily measured using this technique. It consists of measuring the photon emission characteristic of the element potassium, selected by a monochromator and amplified by a photomultiplier. To achieve this, an aliquot fraction of our mineral to be assayed is dissolved by acid etching. Hydrofluoric acid is used to destroy the crystalline silicate lattices and release the various cations involved in the assembly. The resulting solution is injected into the spectrophotometer flame.

This measurement technique is capable of detecting very low concentrations, down to the fraction of a ppm. The measurement range is typically from 1 to 3 ppm, so the etch point and dilution ratio of the residue obtained will be adapted to this sensitivity. This technique enables us to measure the potassium element content with a relative accuracy of the order of 1%, whatever the concentration of our mineral, which can vary between 0.1% (1,000 ppm) and 15%.

– Measuring argon content

Determining the argon content of a very recent sample is extremely delicate, as we have seen how small the quantities to be measured are. It is not possible to collect such quantities completely and measure them directly, so the isotope dilution method is used. Schematically, this involves adding to the unknown quantity of the ^{40}Ar isotope to be determined, a quantity of the same order of magnitude, and yet with a perfectly known quantity (the "spike"), of another isotope known as the "tracer", in fact ^{38}Ar. The ratio of the abundances of the two isotopes is then measured. The mass spectrometer is the instrument for these relative isotope abundance measurements. In reality, things are less straightforward, as there is no such thing as a tracer made entirely of argon-38 (some 40 and 36 are always added), and the mixture always contains varying amounts of argon-36, 38 and 40 of various origins, added to the radiogenic argon in the rock. What is more, the argon released in this manner (and which will have been mixed with the spike as soon as possible, so that any losses affect both components of the mixture equally) will be accompanied by a relatively large quantity of other gases. Most of these will have to be eliminated before being introduced into the mass spectrometer; otherwise, the latter's operation will be disrupted. An argon dosing system therefore comprises the following four components:

– an extraction device consisting essentially of a furnace;

– a calibration device that mixes an exactly known quantity (spike) of tracer with the gases leaving the furnace;

– a purification device which traps almost all the components of the mixture coming from the furnace, other than the noble gases;

– a measuring device comprising the mass spectrometer and its introduction line.

All these devices must be able to be linked together or isolated. It must be possible to pass the gas mixture from the furnace through the purification elements, and then its residue to the mass spectrometer [GOU 75].

The extraction, purification and isotopic analysis processes are as follows:

– Extraction

The argon-40 accumulated by the decay of potassium-40 is extracted from the mineral by vacuum melting using a high-frequency furnace. To do this, an aliquot fraction of the mineral is placed in a crucible inside a vacuum chamber. This chamber is placed at the heart of a solenoid through which a high-frequency current flows and induces a current in the metal structure of the crucible. The polarity of the current alternates at high frequency, creating electronic agitation and thus heating the metal mass of the crucible, which transmits its temperature to the sample by radiation. Temperatures of up to 2,000°C can be reached. Most minerals melt between 800 and 1,200°C, releasing the gases they contain: H_2O, hydrocarbons or CO_2 for the most part, plus the largely diluted argon we are interested in.

– Purification

Argon must therefore be separated from other gases. As a noble gas, argon is chemically inert. It is therefore purified by inducing the chemical reaction of the other gases and combining them with a red-hot metal, usually titanium, which is highly reactive and whose compounds are stable at temperature. This reaction isolates argon and the other rare gases. The most abundant after argon is helium, whose condensation temperature (−269°C, boiling point) is much lower than that of liquid nitrogen (−196°C for N_2). By condensing our purified gas at the temperature of liquid nitrogen, we can pump out the helium. Other rare gases (neon, krypton) are present in such small quantities that they have no influence on the analysis, especially as their chemical inertia has no effect on the operating conditions of the instrument.

– Isotopic analysis

The purified argon is introduced into the mass spectrometer. This measurement is easy for older samples that have accumulated significant amounts of argon. All the argon measured is of a radiogenic nature, and contamination by atmospheric argon is negligible. However, for samples of younger age, the amount of argon accumulated is lower and even more so if they are low in potassium. In this case, the share of contamination becomes paramount. This contamination corresponds to the atmospheric argon occluded in the mineral during its formation, adsorbed on its surface or incorporated into the alteration zones, added during its preparation (grinding, washing) or by the various materials heated during the experimental analysis procedure (Pyrex or quartz glassware, molybdenum crucible for extraction, red-hot titanium foam for purification, etc.). Here, we see the paradox of dating towards recent ages: the younger the sample, the smaller the quantity of radiogenic argon accumulated, and therefore the more diluted the quantity of argon contamination.

Of course, the analyst's first concern (and this is an important part of the progress made in very recent times) will be to reduce the analysis blank and the contamination caused by the experimental procedure. It will also be the analyst's constant concern to use fresh, healthy minerals, whose surface condition and corrosion gulfs will not encourage the adsorption of atmospheric argon. After all, this contamination comes from argon in the atmosphere. As said earlier, argon is no longer a rare gas, because it is produced by the radioactivity of potassium-40. In fact, it is the third most abundant constituent of the air we breathe. Released at depth during mineral transformations (fusion, metamorphism), it migrates to the surface and is released into the atmosphere, where it accumulates. This is why argon-40 is today the most abundant isotope of argon. It is 296 times more abundant than argon-36, the initially most abundant isotope, which does not benefit from any significant radioactive reaction.

With the ratio of argon-40 to argon-36 in the atmosphere set at 295.5, it is now possible to correct for atmospheric contamination. To do this, isotopic measurements are taken of the argon extracted from the sample: 40, 36 and 38. As we have seen, 36 and 38 were produced by nucleosynthesis processes, prior to the individualization of our solar system and the condensation of the planets. They are therefore essentially atmospheric. We can use the measurement of these isotopes to correct for contamination: by multiplying the value of the measured 36 (or 38) by the ratio of this isotope to argon-40 in air (40/36 = 295.5; 40/38 = 1,535), we can deduce the quantity of argon-40 linked to atmospheric contamination in the sample. By simple subtraction from the total argon-40 measured, we obtain the amount of radiogenic argon-40:

$$^{40}\text{Ar}_{\text{total}} = {}^{40}\text{Ar}_{\text{radiogenic}} + {}^{40}\text{Ar}_{\text{contamination}}. \tag{3.16.b}$$

This measurement is made using the isotope 36, as it is five times more abundant than 38 and therefore easier and more accurate to measure. Note that in this correction procedure, the uncertainty of the analytical measurement of 36 is amplified by the multiplication factor 296, an uncertainty that is carried over to the radiogenic argon assay.

3.2.7. Overestimation of K–Ar ages

A K–Ar age will be overestimated if, at $t = 0$, the time of eruption, the sample already contains $^{40}\text{Ar}^*$ not derived from the ^{40}K decay of the mesostasis that crystallized after system closure. Therefore, at $t = 0$, the $^{40}\text{Ar}/^{36}\text{Ar}$ ratio will be greater than 298.56, the value assigned to the atmospheric argon ratio [GUI 17]. This argon, commonly known as "foreign argon" [DAL 69], is of two types. During an eruption, enclaves of ancient rocks from the surrounding basement, or even ancient crystals, may be incorporated into the magma as it rises to the surface. These *xenoliths* are necessarily older than the eruptions that brought them to the surface. This argon is referred to as "inherited argon". These crystals or *lithic fragments* are eliminated from the dated fraction by densitometric and magnetic sorting, leaving only the mesostasis, which is the phase crystallized during surface cooling of the lava.

The second type of argon is "excess" argon. This is the fraction of argon-40 introduced into the sample during crystallization: it does not originate from the in-situ decay of ^{40}K within the sample. The origin of this argon is still poorly understood. This excess argon is generally distributed in the form of fluid or glassy inclusions in the peripheral zones of the rock's constituent grains [KEL 02]. Densitometric and magnetic sorting are ineffective in separating this argon from purely radiogenic in situ argon. All the extracted argon is analyzed by mass spectrometry in a single batch. The tracer-free K–Ar method used [CAS 82] does not separate the $^{40}\text{Ar}^*$ resulting from the decay of ^{40}K mesostasis from the excess ^{40}Ar. This method is described in detail in the work of Sasco [SAS 15].

3.2.8. Description of the $^{40}Ar/^{39}Ar$ dating method

The $^{40}Ar/^{39}Ar$ method makes it possible to distinguish radiogenic argon from excess argon by means of isochrones and age spectra. Samples are subjected to a fast neutron flux in a nuclear reactor, to artificially transform some of the ^{39}K isotopes into ^{39}Ar. The quantity of ^{39}Ar thus generated is proportional to the number of ^{39}K present in the sample, and therefore to the number of ^{40}K (parent atoms), the $^{40}K/^{39}K$ ratio being known at a given instant and therefore at the present time in terrestrial samples. The advantage is that the ratio of radiogenic daughter atoms ($^{40}Ar^*$) versus parent atoms (^{39}Ar proportional to ^{40}K) is obtained by simultaneous measurement using mass spectrometry. Precise knowledge of the yield of ^{39}Ar production from ^{39}K is obtained by reference to standards of known ages, irradiated together with the samples to be dated [GUI 17].

The $^{40}Ar/^{39}Ar$ method provides more comprehensive information on the behavior of the radioisotope clock than the K/Ar method. In the "step-heating" method [TUR 66], the sample is heated progressively in increments of increasing temperature (e.g. increments of 60°C) using an oven or laser. For each step, the argon isotopic composition of the extracted and purified gas is measured with a mass spectrometer. This enables us to calculate an apparent age for each degassing stage of the sample. The final result is an age spectrum. The general appearance of these spectra reveals whether or not the K/Ar clock within the sample has been disturbed. An undisturbed sample that has evolved in a closed system since crystallization will have a constant ratio of $^{40}Ar^*$ to ^{39}Ar isotopes at each temperature step. Therefore, for each temperature step, an identical apparent $^{40}Ar/^{39}Ar$ age will be obtained, to the nearest analytical error. The result of the experiment will be represented in the form of an age spectrum, with each temperature step giving the same age. The age of the sample can then be defined as the plateau age. A plateau is generally considered to be made up of at least three successive stages containing at least 60% of the degassed $^{39}Ar_k$ (index k designates the ^{39}K resulting from the activation of ^{39}K during irradiation), and whose apparent ages are consistent to 95% probability [SHA 05].

For samples containing excess argon, $^{40}Ar/^{39}Ar$ analysis by successive degassing stages highlights the decay of fluid inclusions at low temperatures. This results in over-aged inclusions in the early stages of the age spectrum, and glassy or solid inclusions at high temperatures [ESS 97], which also result in over-aged inclusions in the latter stages of the age spectrum. The age spectrum is then discordant, U-shaped, diagnostic of excess ^{40}Ar. This experimental approach has also demonstrated the migration of excess argon to the grain periphery [HAR 81]. Fluid and glassy inclusions are frequent sources of excess argon and can significantly impact the dating of rocks, particularly those that are young and/or low in potassium. The high argon content of these inclusions is explained by argon's hygromagmatophilic nature (i.e. its strong affinity for fluid or liquid phases). Glassy inclusions from magma rich in excess argon can contain up to 100 times more argon (by weight) than minerals crystallizing from the same magma. In the case of fluid inclusions, this value rises to 10,000 [KEL 02].

The $^{40}Ar/^{39}Ar$ method also makes it possible to process the isotopic data obtained to construct isochrones. The experimental $^{39}Ar/^{40}Ar$ versus $^{36}Ar/^{40}Ar$ values acquired for each temperature step are plotted in an "inverse isochrone" diagram. When the extracted gas is a simple mixture of atmospheric argon and radiogenic argon, the points are aligned along a mixing line or isochrone. The intersection of the isochrone with the $^{36}Ar/^{40}Ar$ axis corresponds to a ^{39}Ar value equal to 0. This deduced and calculated value corresponds to the isotopic signal of a sample devoid of K, which cannot therefore produce purely radiogenic $^{40}Ar*$. Therefore, for a sample devoid of excess $^{40}Ar*$, the isochrone intercepts the $^{36}Ar/^{40}Ar$ axis at an ordinate value of 1/298.56 [LEE 06], hereafter referred to as $(^{36}Ar/^{40}Ar)_i$. Conversely, for a sample containing excess $^{40}Ar*$, $(^{36}Ar/^{40}Ar)_i$ will have a value of less than 1/298.56. The presence or absence of excess argon can thus be determined from the inverse isochron diagram.

3.3. Lake dating using ^{210}Pb, ^{137}Cs and ^7Be radiochronometers

3.3.1. Core drilling system

Among the most widely used methods of *radiochronometric* dating using a *coring system* are the ^{210}Pb, ^{137}Cs and ^7Be radiochronometers. These are ideal for dating soils and sediments.

Figure 3.9(a) shows an example of an interface corer dedicated to sampling 9 cm diameter cores. The length of a core can be modulated according to the user's needs. However, for practical reasons, when the coring machine is lifted into the Zodiac-type transport boat, the length of a core should not exceed 2 meters [DEG 17].

The core sampling principle is described in Figure 3.9b. Core sampling is carried out by gravity (Figure 3.9a), facilitated by the associated weights. Depending on the depth of the core, steps 2 and 3 may need to be repeated.

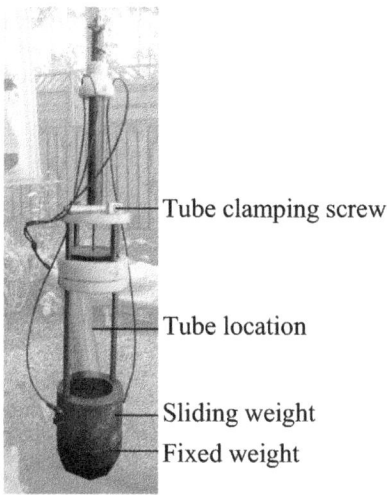

Figure 3.9(a). *Interface corer producing 9 cm diameter cores [DEG 17]*

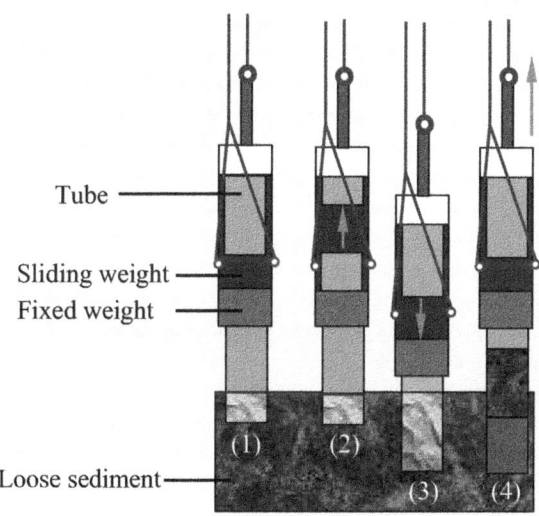

Figure 3.9(b). *Core drilling principle based on an interface corer*

Therefore, through coring, it is possible to take samples of soil or sediment with a view to dating them with lead-210, cesium-137 and beryllium-7. The principles of these three radiochronometers are explained below.

3.3.2. *Lead-210 dating: CFCS, CRS and CIC models*

Lead-210 is one of the decay products of the uranium-238 decay chain. We truncate this decay product as follows [SAK 22]:

$$^{238}_{92}U \xrightarrow[4.468\times10^9\,y]{\alpha} {}^{234}_{90}Th \xrightarrow[24.1\,d]{\beta^-} \dots {}^{226}_{88}Ra \xrightarrow[1600\,y]{\alpha} {}^{222}_{86}Rn \xrightarrow[3.8235\,d]{\alpha} {}^{218}_{84}Po$$

$$[3.17]$$

$$^{218}_{84}Po \xrightarrow[3.05\,min]{\alpha} {}^{214}_{82}Pb \dots {}^{210}_{82}Pb \xrightarrow[22.26\,y]{\beta^-} \dots {}^{210}_{84}Po \xrightarrow[138.38\,d]{\alpha} {}^{206}_{82}Pb\ (stable)$$

As shown in equation [3.17], lead-210 is present in soils and sediments. This progeny of uranium-238 exists under the form of two fractions, one called supported lead-210 ($^{210}Pb_{sup}$) and the other excess lead-210 ($^{210}Pb_{ex}$) [DEG 17]. $^{210}Pb_{sup}$ comes from the radioactive filiation of ^{238}U present in the soil, while $^{210}Pb_{ex}$ comes from radon emanations in the atmosphere, as shown in Figure 3.10.

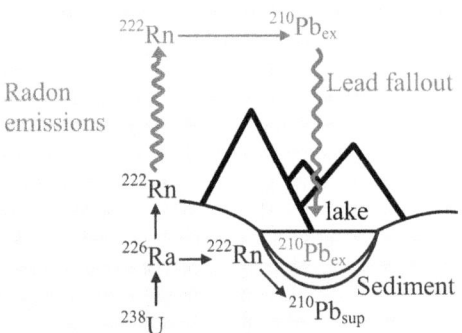

Figure 3.10. *Origins of supported lead-210 ($^{210}Pb_{sup}$) and excess lead-210 ($^{210}Pb_{ex}$) in sediments*

Therefore, total lead-210 ($^{210}Pb_{tot}$) present in soils and sediments is the sum of supported lead and excess lead, i.e.:

$$^{210}Pb_{tot} = {}^{210}Pb_{sup} + {}^{210}Pb_{ex}$$

$$[3.18]$$

For a given radioactive chain, when the period T_1 of the parent nucleus X_1 is very large compared with the periods T_m ($m \geq 2$) of all the child nuclei, a *radioactive equilibrium* called *secular equilibrium* occurs [SAK 22]. According to equation [3.17], the period of radium-226 is $T_1 = 1,600$ years, while that of lead-210 is $T_2 = 22.26$ years. The condition $T_1 \gg T_{m\,=\,2}$ ($T_1/T_2 \approx 72$) can thus be considered satisfied. This allows us to assume that supported lead-210 is in secular equilibrium with radium-226, i.e. ^{210}Pb = ^{226}Ra. Hence, according to [3.18]:

$$^{210}\text{Pb}_{\text{ex}} = {}^{210}\text{Pb}_{\text{tot}} - {}^{226}\text{Ra} \tag{3.19}$$

When the condition $T_1 \gg T_{m=2}$ is satisfied, we show that secular equilibrium is reached at date t_m given by the relation [SAK 22] (see Application 3.3 for demonstration):

$$t_m = \frac{1}{\ln 2} \frac{T_1 T_2}{(T_1 - T_2)} \ln\left(\frac{T_1}{T_2}\right) \tag{3.20}$$

Using [3.20], secular equilibrium between supported lead-210 and radium-226 is reached at t_m equal to:

$$t_m = \frac{1}{\ln 2} \frac{1600 \times 22.26}{(1600 - 22.26)} \ln\left(\frac{1600}{22.26}\right) = 139.22 \text{ years} \tag{3.21}$$

According to [3.21], radioactive equilibrium between supported lead-210 and radium-226 can theoretically only be established in soil or sediment samples older than 139 years. This condition can be met when considering the use of the ^{210}Pb chronometer, which enables the ages of soils and sediments to be determined over a period of up to 150–200 years [APP 78, DEG 17].

In the case of a lake, sediment samples are sealed to prevent radon-222 from escaping from the sample holder and to prevent excess lead-210 from being affected by a new lead contribution [SAN 12; DEG 17].

APPLICATION 3.3.–

Consider a sample of radioactive parent nuclei X_1 with decay constant λ_1. Decay produces intermediate radioactive father nuclei X_m with decay constant λ_m and leads to the stable son nucleus X_n. The decay chain is written as:

$$X_1 \xrightarrow{\lambda_1} X_2 \xrightarrow{\lambda_2} \dots\dots X_{m-1} \xrightarrow{\lambda_{m-1}} X_m \xrightarrow{\lambda_m} X_{m+1} \dots$$

$$\dots X_{n-1} \xrightarrow{\lambda_{n-1}} X_n \text{ (stable)} \tag{3.22}$$

Considering *simple two-body filiation*, show that the *law of accumulation of the filiation product* X_2 is written as:

$$N_2(t) = \frac{\lambda_1}{\lambda_2 - \lambda_1} N_{01}\left(e^{-\lambda_1 t} - e^{-\lambda_2 t}\right)$$

[3.23]

ANSWER.–

Simple two-body filiation corresponds to the simple case of nuclei X_1 ($m = 1$) and X_2 ($m = 2$). The decay chain is written as follows:

$$X_1 \xrightarrow{\lambda_1} X_2 \xrightarrow{\lambda_2} X_3 \text{ (stable)}.$$

[3.24]

During decay, the dN_2 quantity of X_2 nuclei decreases, and also increases by the dN_1 quantity from the decay of the X_1 parent nuclei:

$$dN_2 = -\lambda_2 N_2 dt - dN_1$$

[3.25a]

For the father nucleus:

$$dN_1 = -\lambda_1 N_1 dt$$

[3.25b]

Using [3.25b], equation [3.25a] is written as:

$$dN_2 = -\lambda_2 N_2 dt + \lambda_1 N_1 dt$$

[3.25c]

If N_{01} denotes the initial number of father nuclei X_1, the law of decreasing number N_1 (t) of father nuclei is written according to *Rutherford and Soddy's empirical law*:

$$N_1(t) = N_{01}e^{-\lambda_1 t}$$

[3.25d]

Carrying the solution [3.25d] into relation [3.25c], we obtain:

$$\frac{dN_2(t)}{dt} + \lambda_2 N_2(t) = \lambda_1 N_{01}e^{-\lambda_1 t}$$

[3.26]

The solution to differential equation [3.26] is of the form:

$$N_2(t) = K_1 e^{-\lambda_1 t} + K_2 e^{-\lambda_2 t}$$

[3.27]

In equation [3.27], K_1 and K_2 are constants to be determined.

A particular solution of differential equation [3.27] is obtained for $K_2 = 0$. This leads to the relation:

$$N_2(t) = K_1 e^{-\lambda_1 t} \Rightarrow \frac{dN_2(t)}{dt} = -\lambda_1 K_1 e^{-\lambda_1 t}$$

Using the last equation above, we obtain from [3.27]:

$$-\lambda_1 K_1 e^{-\lambda_1 t} + \lambda_2 K_1 e^{-\lambda_1 t} = \lambda_1 N_{01} e^{-\lambda_1 t}$$

This gives the expression for K_1:

$$K_1 = \frac{\lambda_1}{\lambda_2 - \lambda_1} N_{01} \qquad\qquad [3.28]$$

By replacing K_1 by its expression [3.28] in [3.27], we obtain:

$$N_2(t) = \frac{\lambda_1}{\lambda_2 - \lambda_1} N_{01} e^{-\lambda_1 t} + K_2 e^{-\lambda_2 t} \qquad\qquad [3.29]$$

Taking advantage of the initial conditions for which at $t = 0$, $N_2 (t = 0) = 0$, we determine the expression for K_2 using [3.29]:

$$K_2 = -\frac{\lambda_1}{\lambda_2 - \lambda_1} N_{01} \qquad\qquad [3.30]$$

Carrying [3.28] and [3.30] into [3.27], we obtain the accumulation law [3.23] of the X_2 parent product.

APPLICATION 3.4.–

Using the results of Application 3.2, demonstrate relation [3.20].

ANSWER.–

By definition, the respective activities $A_1(t)$ and $A_2(t)$ of cores X_1 and X_2 are written as:

$$A_1(t) = \lambda_1 N_1(t) \, ; \, A_2(t) = \lambda_2 N_2(t) \qquad\qquad [3.31]$$

The number of father nuclei X_1 decreases exponentially, while that of the intermediate father nucleus X_2 evolves according to the respective laws:

$$A_1(t) = \lambda_1 N_{01} e^{-\lambda_1 t}$$ [3.32a]

$$A_2(t) = \frac{\lambda_1 \lambda_2}{\lambda_2 - \lambda_1} N_{01} \left(e^{-\lambda_1 t} - e^{-\lambda_2 t} \right)$$ [3.32b]

Using [3.32], the activity ratio $A_2(t)/A_1(t)$ is written as:

$$\frac{A_2(t)}{A_1(t)} = \frac{\lambda_2}{\lambda_2 - \lambda_1} e^{\lambda_1 t} \left(e^{-\lambda_1 t} - e^{-\lambda_2 t} \right)$$

The result is:

$$\frac{A_2(t)}{A_1(t)} = \frac{\lambda_2}{\lambda_2 - \lambda_1} \left(1 - e^{(\lambda_1 - \lambda_2)t} \right)$$ [3.33]

Knowing that $T_2 \gg T_1$, then $\lambda_1 \gg \lambda_2$.

When regime equilibrium is reached at date t_m, then $A_2(t_m)/A_1(t_m) = 1$. Using [3.17], we obtain:

$$e^{(\lambda_1 - \lambda_2)t_m} = \frac{\lambda_1}{\lambda_2} \Rightarrow (\lambda_1 - \lambda_2)t_m = \ln\left(\frac{\lambda_1}{\lambda_2} \right)$$ [3.34]

By introducing radioactive periods, the last equality [3.34] gives equation [3.20].

From an experimental point of view, three models can be used to determine the age of a sediment according to assumptions made about the geological characteristics of the lakes. These are the CFCS (Constant Flux and Constant Sedimentation), CRS (Constant Rate of Supply) and CIC (Constant Initial Concentration) models [APP 01, SAN 12, DEG 17].

– CFCS (Constant Flux and Constant Sedimentation) model

The CFCS model is based on the assumptions that the *annual flux density* in $^{210}Pb_{ex}$ noted f (in $Bq \cdot cm^{-2} \cdot a^{-1}$) and the *sedimentation rate r* are constant. This model is generally used for lakes where erosion and precipitation in the water column are stable. Based on these assumptions, the initial $^{210}Pb_{ex}$ concentration is assumed to be constant in each sediment layer i using the equation $C = f/r$. The law of decay of $^{210}Pb_{ex}$ in a sediment layer i is then written as:

$$C_i(t) = C_i(t = 0) e^{-\lambda t}$$ [3.35]

The age of the sediment studied can then be determined within the framework of the CFCS model by taking advantage of two *linear regression* laws using [3.35]. For this, the age t of the sediment is defined by the relations [DEG 17]:

$$t = \frac{m}{r} = \frac{x}{\omega}$$ [3.36]

In relations [3.36]:

– m is the cumulative dry mass per unit area (in $g \cdot cm^{-2}$);

– r is the yearly *sedimentation rate* (in $g \cdot cm^{-2} \cdot y^{-1}$);

– x is the sediment depth (in cm);

– ω represents the *sedimentation rate* (cm $\cdot y^{-1}$).

Using [3.36], equation [3.35] allows us to establish two linear regression laws by plotting the curve $\ln C_i(t)$ as a function of dry mass m or as a function of depth x, i.e. (posing $C_0 = C_i(t = 0)$):

$$\ln C_i(t) = \ln C_0 - \frac{\lambda}{r} \cdot m$$ [3.37]

$$\ln C_i(t) = \ln C_0 - \frac{\lambda}{\omega} \cdot x$$ [3.38]

As indicated by equations [3.37] and [3.38], the CFCS model can only be used if the natural logarithm of the $^{210}Pb_{ex}$ concentration scales linearly either with the cumulative dry mass m [3.37] or with the depth x [3.38].

By determining the leading coefficient a of the regression line [3.37] ($a = \lambda/r$), the age t of the sediment is written as:

$$t = \frac{m}{r} = \frac{m \cdot a}{\lambda}$$ [3.39]

Likewise, by determining the leading coefficient a of the line [3.38] ($a = \lambda/\omega$), the age t of the sediment is written as:

$$t = \frac{x}{\omega} = \frac{x \cdot a}{\lambda}$$ [3.40]

Note that the use of the CFCS model can make it possible to highlight a change in sedimentation along a sampled core. Indeed, a break in the leading coefficient a of the regression lines [3.39] and [3.40] would respectively indicate a change in sedimentation rate r [3.39] or sedimentation speed ω [3.40].

– CRS (Constant Rate of Supply) model

As part of the CRS model, we assume that the $^{210}\text{Pb}_{ex}$ flux f is constant, while the sedimentation rate r can vary over time. It follows that the concentration of $^{210}\text{Pb}_{ex}$ varies with time according to the equation $C = f/r$.

– To put the CRS model into practice, we introduce a new quantity denoted A and called the *inventory* or *accumulation deposit* expressed in $\text{Bq} \cdot \text{cm}^{-2}$ given by the integral [SAN 12, DEG 17]:

$$A_i(t) = \int_m^\infty C_i \, dm$$ [3.41]

As for the concentration C_i of $^{210}\text{Pb}_{ex}$ in a given sedimentary layer i, the inventory A_i decreases exponentially with time, following the law:

$$A_i(t) = \frac{f}{\lambda} e^{-\lambda t}$$ [3.42a]

Let us say (knowing that $C_i = f/r_i$):

$$A_i(t = 0) = \frac{f}{\lambda} = \frac{C_i(t = 0) r_i}{\lambda}$$ [3.42b]

In the last equality [3.42b], r_i is the sedimentation rate associated with sedimentary layer i in which the initial concentration of $^{210}\text{Pb}_{ex}$ is C_i ($t = 0$). Taking into account [3.42b], equation [3.42a] is then written as:

$$A_i(t) = A_i(t = 0) e^{-\lambda t}$$ [3.42c]

Using [3.42c], the age of the sediment is given by the expression:

$$t = \frac{1}{\lambda} \ln\left(\frac{A_i(t=0)}{A_i}\right)$$

[3.43]

APPLICATION 3.5.–

Using [3.35] and the relation $C = f/r$, prove the relation [3.42a]. *Data*: $t = m/r \Rightarrow$ $r_i = dm/dt$.

ANSWER.–

Within the framework of the CRS model, the flux f is constant, while the sedimentation rate r can vary over time. The concentration of $^{210}Pb_{ex}$ varies with time according to the equation $C = f/r$. Using [3.35], we obtain for a layer i of sediment:

$$C_i(t=0) = \frac{f}{r_i} \Rightarrow C_i(t) = \frac{f}{r_i} e^{-\lambda t}$$

[3.44]

Knowing that $dm = r_i dt$ and taking into account [3.44], equation [3.41] becomes:

$$A_i(t) = \int_m^\infty \frac{f}{r_i} e^{-\lambda t} dm = f \int_t^\infty e^{-\lambda t} dt$$

[3.45]

By integrating [3.45], we obtain the relation [3.42a].

– CIC (Constant Initial Concentration) model

The CIC model is based on the hypothesis that the $^{210}Pb_{ex}$ concentration is constant during the formation of sedimentary layer i.

By designating the initial concentration of $^{210}Pb_{ex}$ by $C_0 = C_i(t = 0)$, its concentration $C_i(t)$ decreases with time by following the law [3.19]. By setting $C_i(t=0) = C_0$, we then obtain:

$$C_i(t) = C_0 e^{-\lambda t}$$

[3.46]

Using [3.46], the age of sediment i is written as:

$$t = \frac{1}{\lambda} \ln\left(\frac{C_0}{C_i}\right)$$

[3.47]

Note that in the CFCS model, the $^{210}Pb_{ex}$ flux f and the sedimentation rate r are constant. Within the framework of the CIC model, the flux f_i and the sedimentation rate r_i for a sedimentary layer i can vary with time. Using the last relation [3.42b], we obtain:

$$\frac{f_i}{r_i} = C_i\,(t = 0) = C_0 \qquad\qquad [3.48]$$

Therefore, within the framework of the CIC model, the f_i/r_i is constant in accordance with [3.48].

NOTE.–

Linear regression is a modeling method making it possible to establish a *linear* relationship between a continuous variable called an "explained variable" or dependent variable, and a set of other continuous variables called "explanatory variables" or independent variables. Starting from a cloud of points, a simple linear regression consists of determining a line passing as close as possible to the points of this cloud (or linear adjustment).

3.3.3. *Nuclear tests, Chernobyl accident*

By definition, a *nuclear test* means the explosion of an atomic bomb for experimental purposes. The first test took place on July 16, 1945 in the desert of New Mexico in the United States, three weeks before the bombings of Hiroshima and Nagasaki in Japan. Until 1960, nuclear tests were mainly carried out in the atmosphere. Explosions result in the release and dissemination of radioactive materials into the environment.

The atmospheric nuclear tests carried out by the great powers between 1945 and 1980 constitute to-date the only massive contribution of artificial radionuclides on a planetary scale [REN 15]. Table 3.4 shows the main radionuclides constituting fallout from nuclear weapons tests, listed in ascending order of their radioactive half-lives.

The power of the tests varied greatly, as did the altitude of the shots. The reason is because they were carried out on the ground or on a barge at sea, at the top of a tower, under a balloon or even by release into the upper atmosphere from an airplane [REN 15].

Radionuclide (symbol)	Period	Radionuclide (symbol)	Period
Iodine-131 (^{131}I)	8 days	Iron-55 (^{55}Fe)	2.7 years
Barium-140 (^{140}Ba)	13 days	Antimony-125 (^{125}Sb)	2.8 years
Cesium-141 (^{141}Ce)	33 days	Tritium (^{3}H)	12 years
Ruthenium-103 (^{103}Ru)	39 days	Plutonium-241 (^{241}Pu)	14 years
Strontium-89 (^{89}Sr)	51 days	Strontium-90 (^{90}Sr)	29 years
Yttrium-91 (^{91}Y)	59 days	Cesium-137 (^{137}Ce)	30 years
Zirconium-95 (^{95}Zr)	64 days	Americium-241 (^{241}Am)	433 years
Cesium-144 (^{144}Ce)	280 days	Carbon-14 (^{14}C)	5,700 years
Manganese-54 (^{54}Mn)	310 days	Plutonium-240 (^{240}Pu)	6,600 years
Ruthenium-106 (^{106}Ru)	370 days	Plutonium-239 (^{239}Pu)	24,000 years

Table 3.4. *Main radionuclides constituting the fallout from nuclear weapons tests, classified in ascending order of their radioactive half-lives [REN 15]*

Note that nuclear explosions mainly concern two types of bombs. H bombs (*also called hydrogen bombs, fusion bombs or thermonuclear bombs*) are explosive devices in which the energy produced comes from the fusion of light nuclei, deuterium and tritium (two isotopes of hydrogen) to produce helium. A bombs (*commonly called atomic bombs, fission bombs or nuclear bombs*) are, on the contrary, explosive devices in which energy is produced by nuclear fission with a *critical mass* (48 kg for ^{235}U and 10 kg for ^{239}Pu) of fissile elements, such as uranium-235 or plutonium-239. Note that an H bomb is more powerful and more complex than an A bomb.

NOTE.–

A mass of fissile material is considered critical when it becomes capable of sustaining a chain reaction, taking into account its size, shape, purity and isotopic composition of the material. A numerical measure of criticality is the neutron multiplier coefficient $k = f - l$, where f is the number of neutrons released on average by each atom fission, and l is the average number of neutrons lost, either because they escape from the system or because they are captured by other atoms without producing fission. When $k = 1$, the mass is said to be critical, when $k < 1$, the mass is subcritical, and for $k > 1$, the mass is said to be supercritical. The critical mass of a ball of pure material (unmoderated) in the absence of a reflector is approximately 48 kg for uranium-235 and 10 kg for plutonium-239. If we arrange around the fissile material a coating that reflects part of the neutrons towards it (neutron reflector), we can reduce the critical mass. To prevent the reaction from triggering at any time, the fissile material is given a shape that facilitates the escape of neutrons: separation into two pieces, or hollow ball, therefore of greater surface area. In this way, the critical mass is not reached, and there is therefore no risk of nuclear fission starting without our desire. The explosion is triggered when all the parts of the fissile material are suddenly brought together, in a suitable form, and thus reach a supercritical mass.

Radioactive fallout differs significantly, depending on the type of bomb. Fission reactions produce the entire range of radioelements. Fusion reactions – which occur in H bombs – generate a significant quantity of neutrons which activate the surrounding environments. One of the most important activation products is carbon-14 formed from nitrogen in the air [DEL 23].

Let us consider the case of an A bomb richer in radioactive products. During the explosion, the fission products, the residual uranium-235 and plutonium-239, as well as the construction materials of the device are heated to very high temperatures (greater than 10 million kelvins). A sort of "fireball" forms (*a mushroom characteristic of a nuclear explosion*), expanding and rising into the atmosphere. The mushroom head can remain in the troposphere (0–10 km) or rise into the stratosphere (10–50 km), depending on the power and altitude of the shot. The radioactive particles released then stay in the atmosphere for a few hours or even a few months before falling back to the ground. Three categories of *atmospheric fallout* are then observable: *local fallout, tropospheric fallout* and *stratospheric fallout* (Figure 3.6).

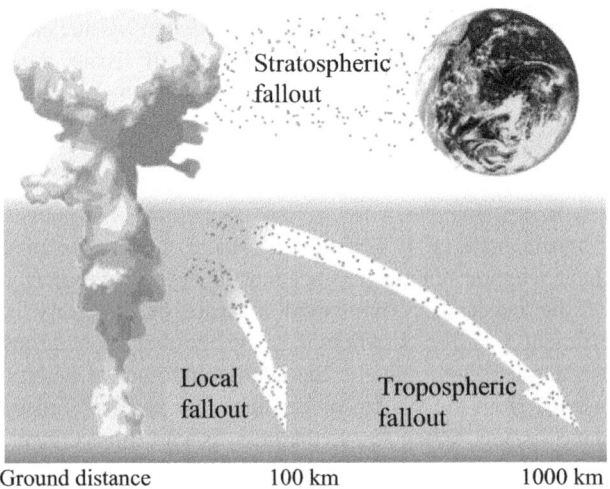

Figure 3.11. *Atmospheric fallout following a nuclear explosion [REN 15]. The troposphere (0–10 km) and the stratosphere (10–50 km) correspond to the first two layers of the atmosphere. The head of the mushroom enters the stratosphere when the power of the shot exceeds 20 kt (20,000 tons). They become essentially stratospheric from 150 kt (150,000 tons) and reach 25 km in height above 1 Mt (1,000,000 tons)*

Local fallout consists of the heaviest debris. These are deposited by gravity within a radius of a few tens to a few hundred kilometers around the shooting site (Figure 3.11).

Tropospheric fallout consists of fission and activation products released into the troposphere. These products stay there for up to 30 days before being deposited on the ground. During this period, they are transported thousands of kilometers by the prevailing winds, which disperse them around the latitude of the shot. As the transport times are relatively short, tropospheric fallout contains the majority of radionuclides, including those with short lives (^{131}I, ^{140}Ba, ^{103}Ru, etc.)

Stratospheric fallout consists of particles released into the stratosphere and which descend by gravity into the troposphere in 2–12 months in polar regions, and in 8–24 months in equatorial regions. This delay leads to good homogenization of the radionuclides and the disappearance of those with short lives. Stratospheric fallout therefore only contains long-lived radionuclides (^{137}Cs, ^{90}Sr, etc.).

Food chain contamination happens when radionuclides are deposited on plant leaves. Among the radionuclides deposited, only iodine-131, cesium-137 and strontium-90 are significantly transferred to consumable parts: roots (potatoes, carrots, etc.), cereal seeds or fruits. These radionuclides are well assimilated by animals after they ingest the contaminated fodder. This is why they also constitute the majority of the radioactivity in milk and meat. Long-lived radionuclides, such as cesium-137, strontium-90 and plutonium isotopes (238, 239, 240, 241) accumulate in soils, thus constituting a secondary source of environmental contamination. Cesium-137 and strontium-90 are transferred to plants by root absorption. This transfer is very weak for plutonium. The leaching of soil by rain contributes to reducing the stock and supplying watercourses. Cesium-137 ($T = 30$ years) and strontium-90 ($T = 29$ years), long-lived radionuclides, are detected in almost all compartments of the food chain due to their high mobility and gradual accumulation in soils. Their long period and mobility maintain a contamination which has only decreased slowly since the mid-1960s [REN 15].

On April 26, 1986, reactor no. 4 at the *Chernobyl* nuclear power plant, which had been in operation since 1983, accidentally exploded. Around 5,000 cases of thyroid cancer are attributable to exposure to radioactive iodine (iodine-131) in people who were children or adolescents at the time of the accident. Employees and the public were exposed to three main types of radionuclides: iodine-131, cesium-134 and cesium-137. Iodine-131 released into the environment is rapidly absorbed by the human thyroid gland. However, iodine-131 has a short half-life (8 days). Children exposed to radioactive iodine receive higher doses than adults, since they have smaller thyroid glands (which grow from around 1.5–20 g from birth to adulthood) and higher metabolic rates. Cesium isotopes have a longer half-life (around 2 years for cesium-134 and 30 years for cesium-137), which increases the risk of long-term exposure through ingestion of contaminated food or water, inhalation of contaminated air or exposure to radionuclides in the soil [COM 22].

VOCABULARY CORNER.–

Metabolism: all the chemical reactions that take place in the body. In particular, energy metabolism encompasses the metabolic pathways and chemical reactions that produce the energy required for cell function.

Let us turn now to the special case of cesium-137 dating.

3.3.4. *Cesium-137 dating*

In the early 1970s, cesium-137 accumulated in soils and became the main contributor to thyroid doses. Thyroid dose calculations are used to assess the risk of developing diseases specific to this organ, which is particularly sensitive to radiation in children [REN 15].

Generally, ^{137}Cs *radiochronometer* dating is used to complement lead-210 dating [DEG 17]. Unlike lead-210, which is a natural decay product of uranium-238, cesium-137 in the environment is exclusively anthropogenic in origin. With the exception of cesium 133, which occurs naturally in the environment, all other cesium isotopes are artificial, produced by nuclear fission reactions. The primary source of ^{137}Cs in the atmosphere is the fission of uranium-235. Under the impact of a thermal neutron, uranium-235 can undergo nuclear fission, producing either xenon-140 and strontium-94, or yttrium-97 and iodine-137. The respective fission equations are as follows:

$$\,_0^1 n + \,_{92}^{235} U \;\rightarrow\; \,_{38}^{94} Sr + \,_{54}^{140} Xe + 2\,_0^1 n + \gamma \qquad [3.49]$$

$$\,_0^1 n + \,_{92}^{235} U \;\rightarrow\; \,_{38}^{97} Y + \,_{54}^{137} I + 2\,_0^1 n \qquad [3.50]$$

Each fission releases two neutrons, which in turn can generate two fissions. The number of fissions can then rapidly become very high, leading to a *chain reaction*. Iodine-137 formed via reaction [3.50] is a β^--emitting fission product, producing cesium-137 according to equation:

$$\,_{54}^{137} I \;\rightarrow\; \,_{55}^{137} Cs + \,_{-1}^0 e + \overline{\nu} \qquad [3.51]$$

Like iodine-137, cesium-137 is a β^- emitter with a half-life of 30.17 years and a *mass activity* of 3.3×10^{12} Bq $\cdot$ g^{-1}. In 94.6% of cases, it transforms into metastable barium-137 (^{137m}Ba) with a decay energy of 512 keV. The ^{137m}Ba then de-excites (by *isomeric transition*) to its fundamental level, emitting a photon γ of energy 661.7 keV with a period of 2.552 minutes. In the remaining 5.4% of cases, cesium-137 decays directly to ^{137}Ba, with a decay energy of 1,174 keV. The transformation chain from ^{137}Cs to ^{137}Ba is as follows:

$$^{137}_{55}Cs \xrightarrow[30.17\,y]{\beta^-} {}^{137m}_{56}Ba \xrightarrow[2.55\,\min]{\gamma} {}^{137}_{56}Ba \qquad [3.52]$$

The global cesium-137 decay scheme is illustrated in Figure 3.12.

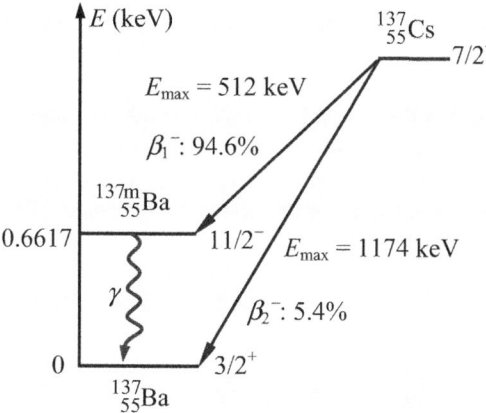

Figure 3.12. *Cesium-137 decay diagram*

Cesium-137 is easily measured in the environment by *gamma spectrometry* via the gamma emission line (0.6617 keV) of its descendant, metastable barium-137.

Atmospheric nuclear testing between 1945 and 1980, and the *Chernobyl nuclear disaster* in 1986, are the second most important source of cesium-137 in the atmosphere. Consequently, cesium was not found in nature before 1945, and even before 1951 [HUB 04]. Generally speaking, four dates relating to unfortunate historical events are used for cesium-137 dating [DEG 17]:

1) 1952: significant start to atmospheric testing of atomic bombs.

2) 1962: 118 atmospheric tests carried out by the USA and the USSR, corresponding to a power of 170 Mt, i.e. over 40% of the total power released between 1945 and 1960. Atmospheric fallout was at its peak.

3) 1963: signing of a treaty against atmospheric nuclear testing between the USA, the USSR and the UK. From this date onwards, these countries began to test underground, and atmospheric fallout diminished.

4) 1986: the Chernobyl accident caused a significant increase in radioactivity in Europe.

More than 75% of the fallout from nuclear testing takes place in the northern hemisphere. Core samples, for example, reveal the unfortunate historical events mentioned above that introduced cesium-137 into the environment.

Therefore, by coring a sediment from Europe at different depths, we obtain a profile of ^{137}Cs activity according to the sampling date (Figure 3.8).

Let us assume that there is no exchange of cesium-137 between the sedimentary layers studied, and that there is no sediment disturbance (i.e. no mixing between sedimentary layers). Based on these assumptions, events relating to atmospheric nuclear testing can be annotated according to measured cesium-137 activity.

Figure 3.13 shows a single ^{137}Cs peak for both the 1973 and 1985 acquisitions. Each of these peaks represents the 1962 nuclear tests carried out by the USA and USSR. The decrease in ^{137}Cs activity from 1962 onwards is justified by the 1963 signing of the treaty against atmospheric nuclear testing between the USA, USSR and UK, which was replaced by underground testing.

Unlike the single-peak acquisitions of 1973 and 1985, the 1996 acquisitions reveal two peaks at depths of around 5 cm and 20 cm. Assuming that the older the sediment, the deeper the layer, the peak at around 5 cm corresponds to atmospheric fallout from the 1986 Chernobyl accident, while the peak at around 20 cm corresponds to the 1962 American and Soviet nuclear tests.

Note that, in the field of *œnology* (a science concerned with the study and understanding of wine), cesium-137 is used to determine the age of wine over several decades and, above all, to authenticate old *vintages* (see vocabulary corner). In addition to deposition on soils and sediments, atmospheric cesium-137 is also deposited on vine leaves and grape bunches. This explains its presence in wine. However, it has not been verified whether this is the case for all wines, or whether transfers from the soil may sometimes occur [HUB 04]. The possibility of *dating wine* is linked to its low radioactivity, mainly from ^{40}K, ^{137}Cs, ^{210}Pb, ^{3}H and ^{14}C.

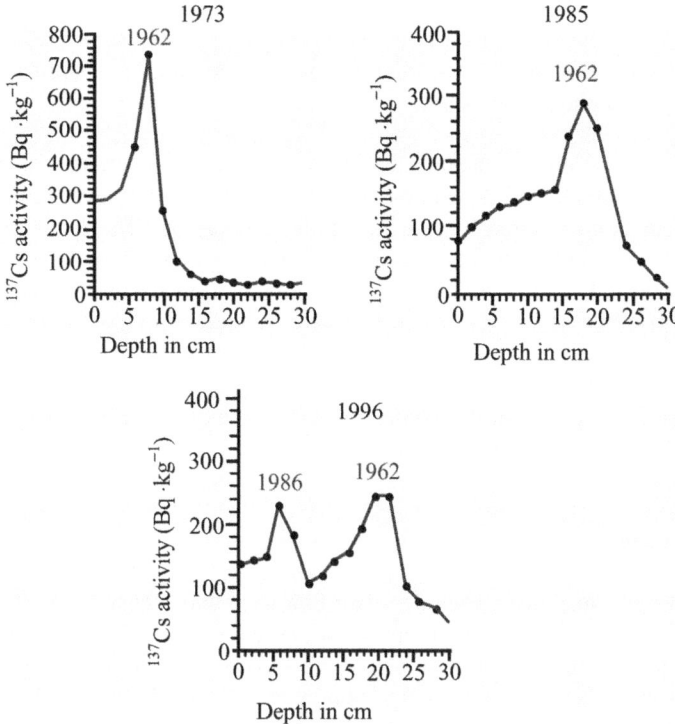

Figure 3.13. *^{137}Cs activity profile as a function of core depth and sampling date*

As shown in Figure 3.13, ^{137}Cs radioactivity can be identified by the emission of the 661.7 keV γ photon resulting from the 137mBa isomeric transition ^{137m}Ba $\rightarrow$^{137}Ba. This photon can be detected through the glass, enabling dating without having to open the bottle, since the presence of ^{137}Cs in a bottle of wine indicates that it was made after 1951.

NOTE.–

New nuclear tests are currently taking place, notably North Korean tests of three A bombs in 2006, 2009 and 2013. The impact of radioactive fallout on the environment can be demonstrated by the use of coring systems.

Vocabulary corner.–

– *Isomeric transition*: a mode of decay in which a nuclear isomer releases all or part of its excitation energy without undergoing transmutation. It is similar to atomic de-excitation with photon emission.

– *Metastable isotope*: an isomer that finds itself in one or more excited states, where it appears stable and de-excites towards the fundamental level of the nuclear isomer via an isomeric transition with γ-photon emission. The letter "m" – for "metastable" – is added to the metastable isotope. Examples include metastable technetium 99, ^{99m}Tc (with 140 keV photon emission, see Chapter 2, Volume 3) and metastable krypton 81, ^{81m}Kr (with 190 keV photon emission, see Chapter 3, Volume 3).

– *Vintage year*: a number designating a year. In oenology, it is the year in which the grapes used to make the wine were harvested.

3.3.5. ^{7}Be dating

Cosmogenic nuclides include beryllium-7 and beryllium-10. Unlike radiocarbon-14, which is formed in the stratosphere (between 12 and 15 km altitude, the stratosphere extends between 10 and 50 km), cosmonuclide ^{7}Be is formed in the troposphere. It is formed by nuclear spallation between a high-energy neutron or proton (greater than 1 MeV) and light atmospheric elements, such as carbon-12, nitrogen-14 and oxygen-16. Possible spallation reactions leading to the formation of ^{7}Be are as follows [PAP 09]:

$$^{12}_{6}C + p \rightarrow \, ^{7}_{4}Be + \, ^{6}_{3}Li \qquad\qquad [3.53a]$$

$$^{12}_{6}C + n \rightarrow \, ^{7}_{4}Be + \, ^{6}_{2}He \qquad\qquad [3.53b]$$

$$^{14}_{7}N + p \rightarrow \, ^{7}_{4}Be + 2\, ^{4}_{2}He \qquad\qquad [3.53c]$$

$$^{14}_{7}N + n \rightarrow \, ^{7}_{4}Be + \, ^{8}_{3}Li \qquad\qquad [3.53d]$$

$$^{16}_{8}O + p \rightarrow \, ^{7}_{4}Be + \, ^{10}_{5}B \qquad\qquad [3.53e]$$

$$^{16}_{8}O + n \rightarrow \, ^{7}_{4}Be + \, ^{10}_{4}Be \qquad\qquad [3.53f]$$

As a radiochronometer, beryllium is used for dating soils and sediments. After its formation in the troposphere via reactions [3.53], ^{7}Be binds to aerosols and is thus present in soils or sediments via atmospheric fallout [MAB 08, SEP 08]. Because of its short half-life of 53.22 days, ^{7}Be is not found at depth, but only in the first 2 cm of core samples [MAB 08, SEP 08]. However, work by Sepulveda et al. [SEP 08] has shown that ^{7}Be can be detected at depths of over 2 cm. This indicates deep particle movement via cracks generated by drought and/or *bioturbation* (vertical soil mixing due to fauna).

From the above, the ^{7}Be radioisotope is indeed a temporal indicator. It can be used to quantify the mixing rate of a soil on a scale of a few months ($\leq$ 6 months [MAB 08]).

The concentration $C\,(x)$ (rather mass activity) of radioberyllium 7 as a function of depth x can be expressed by the relationship [LEH 08] (in this reference, the concentration is denoted by $A\,(x)$):

$$C(x) = C_0 \cdot e^{-x\sqrt{\lambda/D_b}} \, . \tag{3.54}$$

In relation [3.54]:

– $C\,(x)$ is expressed in mBq $\cdot$ g^{-1} and x in cm;

– C_0 is the ^{7}Be concentration at the sediment/water or soil/air interface, C_0 in mBq $\cdot$ g^{-1};

– λ is the decay constant of ^{7}Be expressed in j^{-1};

– D_b represents the bioturbation rate or mixing coefficient in cm$^2\cdot$ j^{-1}.

To determine the bioturbation rate D_b, we can exploit the *linear regression* of ^{7}Be concentration versus sedimentary layer depth x. The result is as follows:

$$D_b = \frac{\ln 2}{T a^2} \, . \tag{3.55}$$

In relation [3.55], *a* s the directing coefficient of the line $\ln C = f\,(x)$ and T is the half-life of ^{7}Be.

Finally, let us establish the equation giving the age of a sediment over a few months using the time indicator ^{7}Be. We know that t is a function of x from [3.40]. To do this, let us equalize [3.46] and [3.54], to obtain:

$$\frac{C(x)}{C_0} = e^{-\lambda t} = e^{-x\sqrt{\lambda/D_b}} \, . \tag{3.56}$$

Using [3.56], we obtain after simplification:

$$t = \frac{x}{\sqrt{\lambda D_b}}.$$ [3.57]

APPLICATION 3.6.–

Demonstrate the expression [3.55] for the bioturbation rate D_b. Estimate D_b for a sedimentary chronology of 5 months, corresponding to a depth of 2 cm. Average 1 month = 30 days.

ANSWER.–

The parameter a is the directing coefficient of the line $C = f(x)$ and T is the half-life of ^{7}Be.

Using [3.56], we obtain:

$$\ln C(x) = \ln C_0 - x \sqrt{\frac{\lambda}{D_b}}$$ [3.58]

Equation [3.58] is a straight line of direction a, with:

$$a = \sqrt{\frac{\lambda}{D_b}} \Rightarrow a^2 = \frac{\ln 2}{TD_b}$$

This gives [3.55].

Using [3.57], we obtain:

$$t^2 = \frac{x^2}{D_b} T \Rightarrow D_b = \frac{T}{t^2} x^2$$ [3.59]

For ^{7}Be $T = 53.22$ days.

Numerically, we find:

$$D_b = \frac{53.22}{(150)^2} (2)^2 = 0.00946 \ \text{cm}^2 \cdot \text{d}^{-1}.$$

VOCABULARY CORNER.–

– *Sediment*: a group of particles suspended in water, atmosphere or ice that has been deposited by gravity, often in successive layers or strata. A sediment is characterized by its nature (physicochemical composition), origin, grain size, species and possible toxicity. The consolidation of sediments is at the origin of the formation of rocky sedimentary layers with varied structures. *Sedimentation* is the set of processes leading to the formation of sediments.

– *Soil*: the surface of the Earth's crust, whether natural or man-made.

3.4. Uranium–thorium or uranium–lead dating

3.4.1. *Method principle*

The uranium-238–thorium-234 or uranium-238–lead-206 radiochronometer dating method is known as the uranium and thorium family imbalance method. This method can be used to determine the age of certain carbonate formations of animal or sedimentary origin. Thorium-234 is a product of the radioactive parentage of uranium-238. This filiation is written as [SAK 22]:

$$^{238}_{92}U \xrightarrow[4.468\times10^9\,y]{\alpha} {}^{234}_{90}Th \xrightarrow[24.1\,d]{\beta^-} {}^{234}_{91}Pa \xrightarrow[1.17\,min]{\beta^-} {}^{234}_{92}U \xrightarrow[2.445\times10^5\,y]{\alpha} {}^{230}_{90}Th$$

$$^{230}_{90}Th \xrightarrow[7.7\times10^4\,y]{\alpha} ... {}^{210}_{84}Po \xrightarrow[138.38\,d]{\alpha} {}^{206}_{82}Pb\ (stable)$$

[3.60]

The truncated decay chain [3.60] shows that uranium-238 decays slowly into thorium-234, with a half-life of 4.5 billion years. It then decays in a few rapid steps (24.1 days and 1.17 minutes) into uranium-234. This isotope, with a half-life of 244,500 years, then disintegrates into thorium-230, which decays via polonium-210 to form stable lead-206. Thorium-230, with a half-life of 77,000 years, is thus well suitable for dating events ranging from 20,000 to 300,000 years ago, provided that the conditions under which the imbalances were created are well known [LAL 79]. The potassium–argon method, mainly applied to the volcanic domain, is not generally applicable to the dating of sedimentary phenomena, as neo-formed potassium minerals are rare [LAL 79]. Therefore, in marine environments, the uranium–thorium ($^{238}U/^{230}Th$) or uranium–lead ($^{238}U/^{206}Pb$) methods are used instead.

In acidic surface waters (pH < 7) such as rivers and lakes, both uranium and thorium are in solution. On the contrary, basic seawater (pH = 8.2) is very rich in dissolved bicarbonate ions. Under these conditions, uranium, which arrives in the form of UO_2^{2+} ions, rapidly complexes to form $UO_2(CO_3)_4^{3-}$, a soluble complex in ocean waters. As for thorium, it exists in the form of Th^{4+} ions, adsorbing to suspended particles and precipitating very rapidly. Therefore, seawater contains uranium 238 (approximately 3 µg · L^{-1}) in secular equilibrium with its descendant ^{234}U (at very low levels: approximately 2 µg uranium per cubic meter of seawater) [CHE 86], but contains practically no thorium ($< 0.7 \times 10^{-4}$ µg · L^{-1}) [LAL 79]. As soon as the river reaches the ocean, thorium precipitates throughout the water mass. A constant flow of thorium-230, separated from its parent uranium-238, reaches the ocean floor. A hard material (coral mineral skeleton, shell, etc.) forming in contact with seawater traps uranium dissolved in the water in its constituent material. Initially, the material therefore contains traces of uranium, but no thorium. However, as ^{234}U decays according to [3.60], thorium-230 accumulates and its content increases, tending towards secular equilibrium. The isotopic ratio $^{234}U/^{230}Th$ therefore provides a measure of elapsed time for ages of around 70,000 years, of the order of magnitude of the half-life of ^{230}Th (77,000 years).

It should also be noted that we can date a sample of mineral material if we know the quantity of lead-206 obtained from the decay of uranium-238 via the nuclear reaction chain [3.60] (see Application 3.7).

APPLICATION 3.7.–

Consider a geological rock containing uranium-238. Let m_U be the mass of remaining uranium and m_{Pb} the mass of lead-206 formed. Determine the age of the rock.

Data: Measurement results: $m_U = 1.03 \times 10^{-12}$ kg; $m_{Pb} = 8.2 \times 10^{-13}$ kg.

ANSWER.–

We use the law of radioactive decay applied to uranium-238. The number of $N_U(t)$ uranium-238 nuclei remaining is then equal to (λ: uranium-238 decay constant):

$$N_U(t) = N_{0U}e^{-\lambda t} = N_{0U}e^{-\ln 2 t/T} \qquad [3.61a]$$

According to equation [3.60], a uranium-238 nucleus decays to give a lead-206 nucleus. Therefore, the number of disintegrated uranium nuclei N_{Ud} is equal to the number N_{Pbf} of lead-206 nuclei formed. The initial number of uranium-238 nuclei is therefore equal to: $N_{0U} = N_U(t) + N_{Ud} = N_U(t) + N_{Pbf}$. Hence, equation [3.61a] is written as:

$$N_U(t) = \left[N_U(t) + N_{Pbf} \right] e^{-\ln 2t/T} .$$ [3.61b]

This gives:

$$e^{\ln 2t/T} = 1 + \frac{N_{Pbf}}{N_U(t)} .$$ [3.61c]

Consider the well-known relationship (N_A: *Avogadro's number*):

$$\frac{m}{M} = \frac{m}{A} = \frac{N}{N_A} \Rightarrow N = \frac{m}{A} N_A .$$ [3.61d]

Using [3.61d], relation [3.61c] is written as:

$$e^{\ln 2t/T} = 1 + \frac{m_{Pbf}}{m_U} \times \frac{A_U}{A_{Pb}} .$$ [3.61c]

The age of the rock is then given by the equation:

$$t = \ln\left(1 + \frac{m_{Pbf}}{m_U} \times \frac{A_U}{A_{Pb}} \right) \times \frac{T}{\ln 2} .$$ [3.61d]

NA: for ^{238}U: $T = 3.368 \times 10^9$ years (equation [3.60]):

$$t = \ln\left(1 + \frac{0.82}{1.03} \times \frac{238}{206} \right) \times \frac{4.468 \times 10^9}{\ln 2} = 4.204 \times 10^9 \text{ years.}$$

3.5. Coral dating

3.5.1. *Uranium-238 decay chain*

Natural uranium contains the isotope 235 (0.7%), whose decay chain produces protactinium-231 (^{231}Pa) with a half-life of 32,700 years. The half-lives of the parent elements are much longer than those of the daughter elements in the decay chain. The final daughter product is stable-lead 207 [SAK 22]:

$$\overset{238}{\underset{92}{U}} \xrightarrow[4.468\times10^{9}\,y]{\alpha} \overset{234}{\underset{90}{Th}} \xrightarrow[24.1\,d]{\beta^{-}} \overset{234}{\underset{91}{Pa}} \xrightarrow[1.17\,min]{\beta^{-}} \overset{234}{\underset{92}{U}} \xrightarrow[2.445\times10^{5}\,y]{\alpha} \overset{230}{\underset{90}{Th}}$$

$$\overset{230}{\underset{90}{Th}} \xrightarrow[7.7\times10^{4}\,y]{\alpha} \overset{226}{\underset{88}{Ra}} \xrightarrow[1600\,y]{\alpha} \overset{222}{\underset{86}{Rn}} \xrightarrow[3.8235\,d]{\alpha} \overset{218}{\underset{84}{Po}} \xrightarrow[3.05\,min]{\alpha} \overset{214}{\underset{82}{Pb}}$$

$$\overset{214}{\underset{82}{Pb}} \xrightarrow[26.8\,min]{\beta^{-}} \overset{214}{\underset{83}{Bi}} \begin{cases} \xrightarrow[19.9\,min]{99,96\,\%\,\beta^{-}} \overset{214}{\underset{84}{Po}} \xrightarrow[163.7\,\mu s]{\alpha} \\[2mm] \xrightarrow[19.9\,min]{0,04\,\%\,\alpha} \overset{210}{\underset{81}{Tl}} \xrightarrow[1.3\,ms]{\beta^{-}} \end{cases} \overset{210}{\underset{82}{Pb}} \xrightarrow[22.26\,y]{\beta^{-}} \overset{210}{\underset{83}{Bi}} \qquad . \ [3.62]$$

$$\overset{210}{\underset{83}{Bi}} \xrightarrow[5.013\,d]{\beta^{-}} \overset{210}{\underset{84}{Po}} \xrightarrow[138.38\,d]{\alpha} \overset{206}{\underset{82}{Pb}}\ (stable)$$

The decay chain [3.62] is often translated into a (Z, N) diagram to represent the natural uranium series (Figure 4.1) [SAK 22].

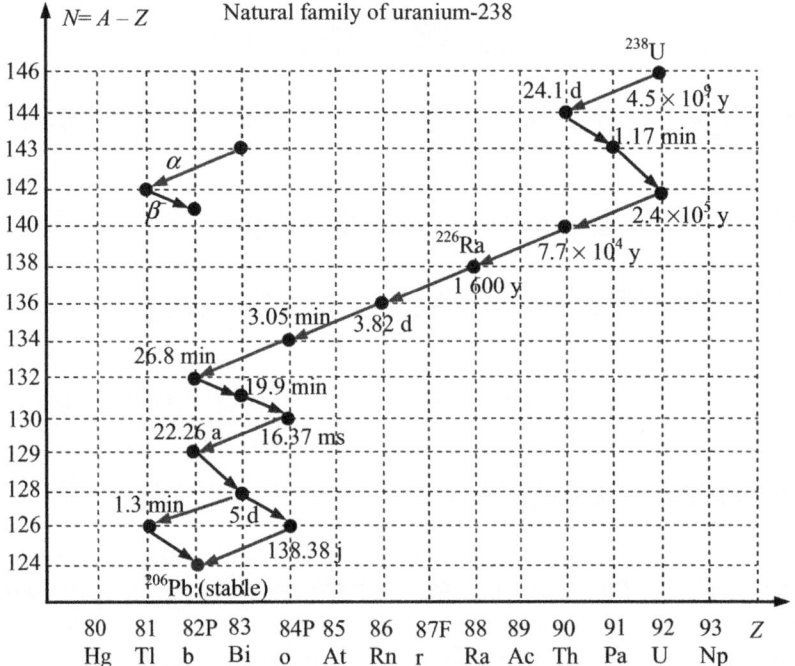

Figure 3.14. *Natural uranium-238 series (often referred to as the uranium-radium series). The stable end product of the series is lead-206*

Like thorium-230, protactinium-231 is also well suited to dating samples (coral, shells, etc.) aged between 20,000 and 300,000 years. From a chronological point of view, the uranium-238–thorium-230 (^{238}U/^{230}Th) radiochronometer and the uranium-238–protactinium-231 (^{235}U/^{231}Pa) radiochronometer were first used in *coral* dating for the determination of the ^{230}Th/^{238}U isotope ratio by Potraz et al. [POT 55] and Barnes et al. [BAR 56] and for the determination of the ^{231}Pa/^{235}U isotope ratio by Sackett [SAC 58, SAC 60].

As pointed out in section 3.4.3, corals represent exceptional biodiversity and can be found in both tropical and cold waters. For fish and other marine animals, corals provide shelter from predators, as well as breeding and nursery grounds for many species. Given the important role they play in the marine environment, coral dating is the subject of intense research. In the following, we describe in detail the principles of coral dating. Emphasis is placed on the experimental aspects, including sampling, mechanical sample preparation, chemical sample preparation, X-ray diffraction analysis and U and Th isotope analysis [CHA 21].

3.5.2. *Sampling, mechanical sample preparation*

– Sampling

In order to date corals, it is necessary to collect sufficient carbonate material in the field. This can be done by hammer sampling or, more commonly, by coral drilling. Before drilling any coral, it is necessary to consider the fossil species present on the study site. Indeed, it is preferable to drill for species with a substantial calcareous skeleton, rather than one with numerous large lodges. The latter are much more prone to weathering, as water infiltrates fossil logs more easily [CHA 21].

– Mechanical sample preparation

Once the corals have been sampled, they need to be cleaned mechanically with a micro-drill (e.g. a "Dremel 3000"), to select only the aragonite pieces. If it is not possible to distinguish between aragonite and calcite with the naked eye, petrographic observations can be made on thin sections of the samples. The main difference between aragonite and calcite lies in their crystallographic structures: aragonite has an orthorhombic crystal system, whereas calcite crystallizes in a trigonal crystal system with a rhombohedral lattice. Prior to cleaning with a micro-drill, coral cores are cut into sections about 1 or 2 cm thick. During mechanical cleaning, we use very fine bits to dig inside the coral lodges. The tips are carefully selected according to the size of the loges, which vary according to coral species. It is very difficult to keep corals intact using this technique, as the calcareous skeleton can be fragmented by the vibrations of the micro-drill and the lack of material in the logs. We then recover all the small, fragmented aragonitic pieces and clean them with sanding bits to flatten the selected surfaces. This removes any calcite layers stuck to the aragonitic limestone skeleton. It is essential not to prolong this stage, i.e. not to polish for a certain length of time, as the micro-drill tip will heat up and may promote the transformation of aragonite into calcite. For further preparation, it is necessary to select around 10 g of aragonitic pieces.

3.5.3. *Chemical preparation of samples, X-ray diffraction analysis*

– Chemical preparation of samples

After the mechanical cleaning described above, several beakers are prepared according to the number of samples selected. These beakers are thoroughly cleaned with Milli-Q (MQ) water and the samples placed in the beakers. The beakers are then filled with MQ water to completely submerge the coral samples, and are then placed in an ultrasonic bath for 20 minutes. This step removes any residual dust stuck to the selected aragonitic pieces. Once the 20 minutes have elapsed, the beakers are emptied and rinsed with MQ water until clear. HCl 0.1 N is then added to the beakers until the samples are covered. The beakers are again placed in an ultrasonic bath for 20 minutes. This step dissolves the calcite still present on the walls of the selected parts. This leaching procedure does not alter the samples or affect the dating results. The samples are then rinsed with MQ water, the beakers filled with MQ water and placed once again in the ultrasonic bath for 20 minutes to clean them one last time. Finally, the beakers are placed in an oven at around 50–60°C overnight.

Once the samples are dry, they are ground using a crusher or agate mortar. The collected powders are then sieved. Approximately 1 or 2 g of grain fractions smaller than 100 μm are recovered. These powders are analyzed by XRD (X-ray diffraction) to determine the percentage of aragonite and calcite in the samples.

Note that the Milli-Q® Reference system provides ultrapure water suitable for a multitude of research sectors.

– X-ray diffraction (XRD) analysis

Each mineral is made up of molecules that are organized in a specific way, forming what is known as a crystal lattice. When a sample is analyzed with an XRD, the radiation is diffracted by the lattice planes of the different minerals in the sample. In other words, the radiation is deflected at one or more angles characteristic of the crystals present in the sample. This method makes it possible not only to identify the minerals present (in our case, aragonite and calcite), but also to assess the proportions of each mineralogical phase.

U and Th isotopes are analyzed using multi-collector inductively coupled plasma mass spectrometry.

3.5.4. *Coral dating using the* $^{238}U/^{230}Th$ *and* $^{235}U/^{231}Pa$ *methods*

As an illustrative example, let us consider a coral forming its calcium carbonate skeleton currently in the sea and neglect for the moment the imbalance between uranium-233 and uranium-238 [LAL 79]. In the aragonite crystals forming this skeleton, uranium from seawater, which is associated with bicarbonate ions in the form of the $UO_2(CO_3)_3^{3-}$ complex, is introduced into the crystal lattice. However, neither thorium-230 nor protactinium-231 is introduced into the crystal. For a coral of "zero" age at the time of analysis, we should find some activity due to uranium and no activity due to thorium or protactinium. Once the skeleton has formed, thorium-230 and protactinium-231 will form at the expense of uranium, and the activity ratios $^{230}Th/^{238}U$ and $^{231}Pa/^{235}U$ will be a function of time only, according to the relationships:

$$^{230}\text{Th}/^{238}\text{U} = 1 - e^{-\lambda t}; \; \lambda = \lambda_{230\text{Th}} \hspace{3cm} [3.63a]$$

$$^{231}\text{Pa}/^{235}\text{U} = 1 - e^{-\lambda t}; \; \lambda = \lambda_{231\text{Pa}} \hspace{3cm} [3.63b]$$

The evolution over time of the isotope ratios $^{230}Th/^{238}U$ and $^{231}Pa/^{235}U$ for samples not initially containing thorium-230 and protactinium-231 is shown in Figure 3.15.

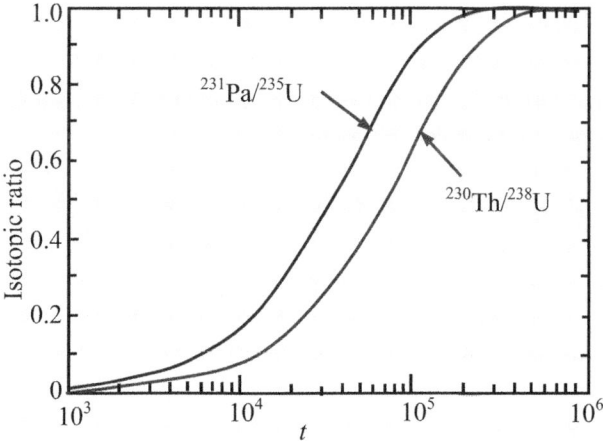

Figure 3.15. *Time evolution of isotopic ratios* $^{230}Th/^{238}U$ *and* $^{231}Pa/^{235}U$ *for samples not initially containing thorium-230 and protactinium-231*

Note that the curves shown in Figure 3.15 are obtained by neglecting the decay with time of uranium-238 and uranium-235. This approximation is justified by the fact that the half-lives of the daughter nuclei ^{230}Th (77,000 years) and ^{231}Pa (32,700 years) are very short, compared with those of their respective fathers ^{238}U (3.5 billion years) and ^{235}U (7 billion years).

3.5.5. Coral dating using the $^{233}U/^{230}Th$ method

In 1955, Cherdyntsev et al. [CHE 55] discovered that uranium-233 was rarely in equilibrium with its ancestor uranium-238. Thurber [THU 62, THU 63] measured excess uranium-233 in corals and seawater. Since then, it has become apparent that the ^{238}U/^{230}Th radiochronometer method is too simplistic and that the determination of the ^{233}U/^{230}Th isotope ratio could be used to date corals. This would make it possible to determine ages closer to reality than those deduced from the ^{230}Th/^{238}U ratio. However, to use the ^{233}U/^{230}Th radiochronometer, the decay of uranium-233 can no longer be neglected, since its half-life of 233,500 years is of the same order of magnitude as the ages measured by the uranium-233–thorium-230 radiochronometer (^{233}U/^{230}Th).

To date corals more accurately, we take into account the fact that as soon as uranium is fixed in the carbonate crystal, the excess uranium-233 decreases to reach equilibrium with uranium-238 at about 1,000,000 years, according to equation [LAL 79]:

$$\frac{^{238}U}{^{233}U} = 1 + \left[\left(\frac{^{238}U}{^{233}U} \right)_0 - 1 \right] e^{-\lambda_{233}t} \qquad [3.64]$$

In equation [3.64], $(^{233}U/^{238}U)_0$ denotes the isotope ratio measured at the initial instant in the coral sample under consideration.

Taking into account the activity of thorium-230, the evolution with time of the isotopic ratio ^{233}Th/^{233}U is governed by the law [LAL 79, SHE 85]:

$$\frac{^{230}Th}{^{233}U} = \frac{^{238}U}{^{233}U}\left(1 - e^{\lambda_{230}t}\right) + \left(1 - \frac{^{238}U}{^{233}U}\right)\left(\frac{\lambda_{230}}{\lambda_{230} - \lambda_{233}}\right)\left(1 - e^{-(\lambda_{230}-\lambda_{233})t}\right) \qquad [3.65]$$

NOTE.–

Coral is a marine animal, often living in colonies, with an orifice surrounded by tentacles that enable it to capture prey for food. It builds a hard, calcareous external skeleton that contributes to the construction of a coral reef. The tree-like structure created by corals is called a "polypier". Corals represent exceptional biodiversity, and can be found in both tropical and cold waters. For fish and other marine animals, corals provide shelter from predators, as well as breeding and nursery grounds for many species. They are often found in waters with low phytoplankton content, which is the source of the marine food chain, and offer a veritable oasis of life in the middle of an oceanic desert. They also provide an ideal natural barrier against cyclones, storms and erosion, as they absorb the power of the waves. Coral reefs are veritable oases of life in the middle of the oceans. They cover just 0.2% of the surface of the seas, but are home to over 25% of the world's marine biodiversity, i.e. almost 60,000 species described to-date. As one of the most diverse ecosystems on the planet, they provide a direct or indirect livelihood for 500 million people, including 30 million fishermen. Reefs are suffering the impacts of human activities, as well as those of climate change (warming and acidification of the ocean). Their global decline is alerting decision-makers to the need to propose protection strategies that reconcile biodiversity conservation with sustainable fishing.

Sources: https://www.oceano.org/ocean-en-questions/quel-est-le-role-du-corail/, https://www.futura-sciences.com/planete/definitions/animaux-corail-15718/, https://www.ird.fr/recifs-coralliens-concilier-conservation-de-la-biodiversite-et-peche-durable.

VOCABULARY CORNER.–

– *Aragonite*: mineral composed of calcium carbonate $CaCO_3$ (with traces of Sr, Pb and Zn).

– *Cyclone*: any atmospheric disturbance characterized by strong lows, torrential rain and winds in excess of 118 km/h.

– *Storm*: a violent atmospheric disturbance; fast, gusty winds, often accompanied by thunderstorms.

– *Erosion*: the process of degradation and transformation of landforms (strong vertical variation of a solid surface).

APPLICATION 3.8.–

Here we consider a sample that may consist of a piece of stalagmite. The age of the sample is determined experimentally using the $^{230}Th/^{233}U$ method. The results of the measured isotope ratios are given [SHE 85]:

^{230}Th/^{233}U: 0.333 ± 0.013; ^{233}U/^{238}U:1.035 ± 0.010; measured age: $(61 \pm 3) \times 10^3$ years.

Show that equation [3.65] can be used to date stalagmites. Determine the accuracy of the calculations. Conclude.

Remember that stalagmites in caves are made up of limestone dissolved in drops of water falling on the floor.

ANSWER.–

From the preceding text, we derive the half-life of thorium-230 ($T = 77{,}000$ years) and uranium-233 ($T = 233{,}500$ years). The decay constants are respectively equal to:

$$\lambda_0 = 9.0 \times 10^{-6} \text{ years}^{-1} \text{ and } \lambda_1 = 2.8 \times 10^{-6} \text{ years}^{-1} \Rightarrow (\lambda_0 - \lambda_1) = 6.2 \times 10^{-6} \text{ years}^{-1} \quad [3.66]$$

Using [3.65], let us find the value of the ^{230}Th/^{233}U isotope ratio to compare with the experimental value. We obtain (age: 0.061×10^6 years; ^{238}U/^{233}U: 0.957):

$$^{230}\text{Th}/^{233}\text{U} = 0.957\,(1 - e^{-9 \times 0.061}) + (1 - 0.957)(9/6.2)(1 - e^{-6.2 \times 0.061}) = 0.323 \quad [3.67]$$

The result [3.67] is close to the experimental value ^{230}Th/^{233}U $= 0.333$.

The calculation accuracy is equal to $|0.333 - 0.323|/0.333 = 2.3\%$.

Conclusion: equation [3.65] can be used to estimate the age of a stalagmite sample to a good approximation.

APPLICATION 3.9.–

Answer the same question in Application 3.8. The conclusion is not necessary.

The following experimental results are given [SHE 85]:

^{230}Th/^{233}U: 0.310 ± 0.020; ^{233}U/^{238}U:1.038 ± 0.025; measured age: $(57 \pm 3) \times 10^3$ years.

^{230}Th/^{233}U: 0.603 ± 0.023; ^{233}U/^{238}U:1.019 ± 0.018; measured age: $(100 \pm 7) \times 10^3$ years.

ANSWER.–

Using [3.65] and [3.66], we obtain:

– age: 0.057×10^6 years; $^{238}U/^{233}U$: 0.963:

$$^{230}Th/^{233}U = 0.963 \ (1 - e^{-9 \times 0.057}) + (1 - 0.963)(9/6.2)(1 - e^{-6.2 \times 0.057}) = 0.302 \qquad [3.68]$$

– age: 0.100×10^6 years; $^{238}U/^{233}U$: 0.981:

$$^{230}Th/^{233}U = 0.981 \ (1 - e^{-9 \times 0.1}) + (1 - 0.981)(9/6.2)(1 - e^{-6.2 \times 0.1}) = 0.595 \qquad [3.69]$$

Results [3.68] and [3.69] are similarly close to the respective experimental values $^{230}Th/^{233}U = 0.310$ and $^{230}Th/^{233}U = 0.603$. The calculation accuracies are 1.9% and 1.5% respectively.

3.5.6. *Dating corals and speleothems using $^{234}U/^{238}U$ and $^{230}Th/^{238}U$ methods*

Speleothems, more commonly known as concretions, are mineral deposits precipitated in natural underground cavities (caves, chasms, etc.). These speleothems generally consist of calcium carbonate (calcite, aragonite) or calcium sulfate (gypsum) transported in solution in percolating water (passage of a fluid through a more or less permeable porous or fissured medium). In contact with the warmer air of a cavity, the solution, while trickling or spurting (then possibly transported as an aerosol), may precipitate as a result of water evaporation and/or CO_2 discharge.

In archaeology, this method is most frequently used to date speleothems. The principle is based on the differential solubility properties of uranium and its radioelements. Therefore, to summarize a rather complex and well-known chemistry [ROS 66, ROS 82, GAS 82], we consider that uranium in its highest oxidation state (+VI) forms highly soluble ions, whereas thorium ions (+IV state) are virtually insoluble under natural pH conditions [GUI 18]. Soil leaching by runoff and infiltration water alters the minerals, which in turn release elements. The most soluble of these are carried away in the aqueous phase. In karstic environments, seepage water is loaded with calcium and carbonate ions. Under certain pH and carbon dioxide conditions, calcium carbonates precipitate out of cavities, trapping uranium ions present in the water. The calcium carbonate formed therefore initially contains trace amounts of uranium ions (^{238}U, ^{235}U, ^{234}U), but no thorium. Over time, the ^{230}Th produced by ^{234}U accumulates. At the same time, the decays of ^{234}U are offset by those of ^{238}U, the longer-lived parent isotope. The ratio $^{230}Th/^{234}U$ goes from 0 at the origin of time and asymptotically tends towards 1. This dating method assumes that the material does not alter over time and is not subject to uranium incorporation or depletion after formation, as is frequently the case with bones. This is called a closed environment, and one of the problems with the method concerns the integrity of the material. This is why the qualification of a dating result necessarily requires a detailed characterization of the dated objects (microscopic observations, study of composition, structure and texture, etc.) [GUI 18].

In 1977, the measurement technique used for U-Th dating was almost exclusively alpha spectrometry. When mass spectrometers coupled to thermo-ionization of sample targets became available in the 1980s, measurement quality took a quantum leap. Accuracy and sensitivity increased by at least an order of magnitude (from an uncertainty of the order of 5% or more with alpha spectrometry, to less than 1% with TIMS: *thermo-ionization mass spectrometry*). This also paved the way for ever finer applications of the method, enabling chronologies to be established with much better chronological resolution [GUI 18].

The principle of speleothems using the ^{234}U/^{230}Th method is based on the difference in solubility between soluble uranium and thorium, which is only slightly soluble in natural waters. In theory, the calcite (or aragonite) that forms speleothems contains no ^{230}Th at the time of precipitation, which accumulates over time through the decay of ^{234}U. The inverse of the ratio of these two elements is therefore proportional to the age of the concretion. The principles of coral and speleothem dating using the ^{234}U/^{238}U and ^{230}Th/^{238}U clocks are outlined below.

In an undisturbed system, after several million years, the daughter elements naturally reach a state of radioactive or secular equilibrium with the parent element [JAF 71, STI 09]. In nuclear physics, secular equilibrium is reached when the rate of production of a radioactive isotope is equal to its rate of decay [CHA 21]. However, natural processes can disrupt the state of isotopic equilibrium, and the return to equilibrium makes it possible to quantify time and thus date the moment of chemical or physical fractionation [BOU 03]. This is the key to U-series disequilibrium dating methods [SCH 08]. In theory, U-series radiochronology can provide precise ages for the last 500,000 to 600,000 years [CHE 13]. In the case of corals, disequilibrium begins when there are no more skeletal deposits (i.e. when the coral is no longer growing) or when the coral dies.

A state of disequilibrium in the ^{238}U decay chain can result either from elemental fractionation of Th from U, or from isotopic fractionation between ^{234}U and ^{238}U. Elemental fractionation results from the different geochemical behavior of U and Th. Uranium exists mainly in two oxidation states in nature (U^{4+} and U^{6+}), and at the Earth's surface, it is dominant in its soluble U^{6+} form. It is soluble as uranyl ion $(UO_2)^{2+}$ and in various forms of uranyl carbonate [SCH 08].

However, in a reducing environment, it appears mainly in the U^{4+} state, where it is insoluble and therefore much less mobile than U^{6+}. In contrast, Th, which is present in terrestrial materials mainly in the 4+ oxidation state, is insoluble in natural waters and, consequently, under natural conditions, is generally transported in minerals or adsorbed on particles [SCH 08].

As a result, groundwater, river water and seawater contain dissolved U but essentially no Th. Therefore, unlike U, Th is not incorporated into secondary carbonates during their formation, resulting in an initial Th/U imbalance. The fractionation of ^{234}U and ^{238}U is produced by the α recoil effect. α decay results in the emission of a He nucleus with finite kinetic energy [SCH 08]. This has a dual effect:

– The daughter nuclide is slightly displaced from its parent's site of origin and thus either directly ejected into an adjacent phase, or more easily eliminated thereafter.

– The crystal lattice is damaged by the α particle along the trajectory [BOU 03]. As a result, the daughter nuclide is then more easily mobilized than its parent, for example, during weathering processes. As a result, $^{234}U/^{238}U$ isotope ratios in groundwater and river water are generally above equilibrium [CHA 03, POR 03].

Various studies have shown that the U/Ca ratio of corals is controlled by the absolute U concentration of seawater [CRO 83, MIN 95, SHE 95, SWA 82]. Isotope fractionation between ^{234}U and ^{238}U does not occur during coral growth [COB 03], which is demonstrated by the good agreement of the mean $\delta^{234}U$ $(^{234}U/^{238}U_{initial})$ value measured on modern corals with the seawater value, i.e. $\delta^{234}U = 146.6 \pm 1.4‰$ [DEL 02] or $^{234}U/^{238}U_{initial} = 1.1468 \pm 0.004$ (0.008 for the less stringent cases [AND 10]).

Molar values of $^{232}Th/^{238}U$ in young corals are slightly lower than in seawater [CHE 86]. These authors showed that U and Th are not fractionated significantly during aragonite formation and are incorporated into the coral skeleton in their proportions relative to seawater. Consequently, the low activity ratio $(^{230}Th/^{238}U)$ in seawater leads to an initial activity ratio close to zero $(^{230}Th/^{238}U)$ in corals, the basic postulate for U-series dating [EDW 86, SCH 08].

Unlike U, Th is not incorporated into secondary carbonates. As a result, the initial activity of ^{230}Th, $(^{230}Th)_0$, is zero. If we further assume that the decay system remains closed after deposition (i.e. both U and Th are neither lost nor gained) and take into account that the half-life of ^{238}U is much longer than that of ^{234}U and ^{230}Th

(Figure 4.1), the evolution of the isotopic ratios ($^{234}U/^{238}U$) and ($^{230}Th/^{238}U$) are given by the following equations [SCH 08]:

$$\left(\frac{^{234}U}{^{238}U}\right)(t) = \left[\left(\frac{^{234}U}{^{238}U}\right)_0 - 1\right] e^{-\lambda_{234}t} + 1.$$

[3.66]

$$\left(\frac{^{230}Th}{^{238}U}\right)(t) = \left(1 - e^{\lambda_{230}t}\right) + \left[\left(\frac{^{234}U}{^{238}U}\right)(t) - 1\right] \frac{\lambda_{230}}{\lambda_{230} - \lambda_{234}} \left(1 - e^{-(\lambda_{230}-\lambda_{234})t}\right)$$

[3.67]

In the above equations, ($^{234}U/^{238}U$)$_0$ is the initial activity ratio ($^{234}U/^{238}U$). The most recent values of the decay constants are equal to [CHE 00]:

$$\lambda_{230} = 9.1577 \times 10^{-6}\ a^{-1}\ \text{and}\ \lambda_{234} = 2.8263 \times 10^{-6} a^{-1}.$$

[3.68]

Note that equations [3.66] and [3.67] are only valid if two basic assumptions are met [SCH 08]:

– no initial ^{230}Th;

– the system remains closed after deposition. If one of these assumptions is not met, the resulting ^{230}Th/U age could be considerably biased and therefore meaningless. However, if both assumptions are met, the measurement of ($^{234}U/^{238}U$) and ($^{230}Th/^{238}U$) can be used to calculate the time elapsed since sample formation.

Furthermore, equation [3.66] alone cannot be used to calculate ages, since the initial isotopic ratio ($^{234}U/^{238}U$)$_0$ is generally unknown. Fortunately, however, equation [3.67] requires only the measurable quantities ($^{230}Th/^{238}U$) and ($^{234}U/^{238}U$) of the sample to resolve the age. Furthermore, equation [3.67] cannot be solved analytically for a given time t. We then resort to graphical solving. In such cases, graphical or numerical resolution is used to calculate the age of a given sample. The corresponding age error can be calculated using numerical methods, such as Monte-Carlo simulation [LUD 03]. A common way of visualizing the time evolution of activity ratios is to plot ($^{234}U/^{238}U$) as a function of ($^{230}Th/^{238}U$). However, when ($^{234}U/^{238}U$)$_0$ is known, equation [3.66] provides the ratio ($^{234}U/^{238}U$) at time t. The value of the ratio ($^{230}Th/^{238}U$) is then deduced for the same instant, using equation [3.66] and the values [3.68] of the decay constants.

Figure 3.16 shows the temporal evolution of ($^{230}Th/^{238}U$) and ($^{234}U/^{238}U$) for two different values of the initial isotopic ratio ($^{234}U/^{238}U$)$_0$.

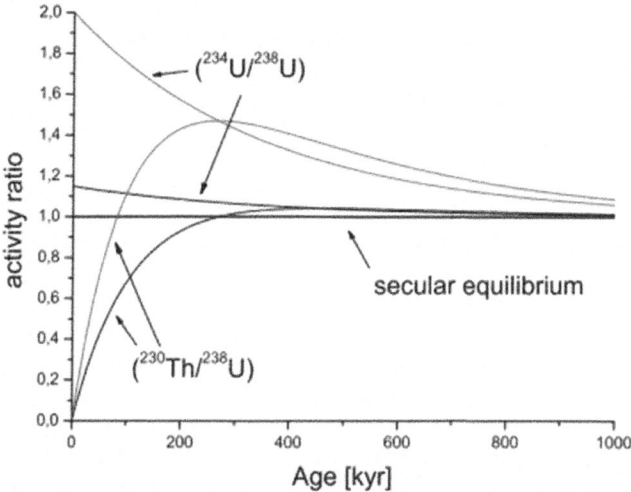

Figure 3.16. *Time evolution of the isotope ratios $(^{234}U/^{238}U)$ and $(^{230}Th/^{238}U)$ under closed system conditions with no initial Th. Two initial values of the $(^{234}U/^{238}U)$ ratio are shown: $(^{234}U/^{238}U)_0. = 1.15$ (black line) and $(^{234}U/^{238}U)_0 = 2.0$ (red line) [SCH 08]*

As shown in Figure 3.16, the evolution of the isotopic ratio $(^{230}Th/^{238}U)$ depends on that of $(^{234}U/^{238}U)$ [3.67], which in turn depends on $(^{234}U/^{238}U)_0$ [3.66]. The limit of the dating range is reached when secular equilibrium is established. Clearly, the dating range depends on the initial value of $(^{234}U/^{238}U)$.

Marine samples generally have an initial value of $(^{234}U/^{238}U)_0 = 1.15$ (black line, Figure 3.16), with a dating range of around 600,000 years [EDW 03]. For other samples, such as speleothems, higher values of $(^{234}U/^{238}U)_0$ can be found, and older samples may still be datable [SCH 08].

Experimentally, two methods are commonly used for U and Th isotope measurements: alpha spectrometry and mass spectrometry. Both techniques require prior chemical separation and purification of U and Th from the $CaCO_3$ matrix. Technical advances in thermal ionization mass spectrometry (TIMS: acronym for Thermal Ionization Mass Spectrometry) in the 1980s led to a significant improvement in accuracy and detection limits [EDW 87]. In the last 10–15 years, further improvements in plasma source mass spectrometry, in particular, the advent of multi-collector inductively coupled plasma mass spectrometry (MC-ICPMS: acronym for Multi Collector Inductively Coupled Plasma Mass Spectrometry), have led to a further evolution towards this technique. The TIMS and MC-ICPMS techniques enable high-precision measurements of U and Th isotopes. A brief description of these two techniques is given below [SCH 08]:

The basic concept of mass spectrometry is:

– isotope ionization;

– acceleration of charged particles using a high potential difference;

– separating beams of charged particles with different mass/charge ratios in a magnetic field.

The intensities of the separate ion beams must then be detected to obtain the relative abundance. A mass spectrometer comprises three main parts: the ion source, the mass analyzer and the detection system.

A TIMS is a solid-state mass spectrometer. Purified U and Th solutions are placed on a filament, for example, made of rhenium, which is placed in the mass spectrometer's ion source. The sample chamber is evacuated, and the filament is heated by an electric current. Atoms are thermally released from the filament, and a certain proportion (< 1%) is also ionized. The ions are accelerated by a high potential in the ion source, and the ion beam is focused into the mass analyzer section. The mass analyzer of a TIMS is a magnetic field. Typically, a U or Th isotopic measurement takes a few hours. The total quantity of U required for an analysis is of the order of tens to hundreds of nanograms, and accuracies of a few per thousand can be obtained.

An MC-ICPMS is a plasma source mass spectrometer. The ion source consists of a torch in which an Ar plasma is generated by a radio-frequency (RF) coil. The purified U and Th sample is dissolved in a weak acid, and these solutions are nebulized in a spray chamber and mixed with the gaseous Ar sample. The mixture is then injected into the plasma, which has a temperature of around 8,000 K and is almost completely ionized (> 90%). Ions are transferred from the atmospheric pressure plasma to the high-vacuum mass spectrometer via the interface region, where over 90% of the ions are lost. Inside the mass spectrometer, an extraction voltage accelerates the ions via electric lenses in the mass analyzer section.

The detection system is largely the same for TIMS and MC-ICPMS. It is generally a combination of Faraday-cut detectors and ion counters. Ion beam intensities above 1 mV can be measured on Faraday cups. Smaller ion beams are detected using ion counters, such as secondary electron multipliers or Daly detectors.

NOTE.–

A Daly detector is a gas-phase ion detector consisting of a metal "door handle", a scintillator (phosphor screen) and a photomultiplier. This was invented by British physicist Norman Richard Daly (1928–2014). Daly detectors are generally used in mass spectrometers.

3.6. File on dating archaeological objects

The following texts provide the interested reader with an overview of the dating of archaeological objects [RIC 21] and some useful references.

3.6.1. *General points*

Since the 1950s [MAY 19], the number of museums has multiplied. Their collections are overflowing with objects of all kinds: art, archaeological, historical, ethnological, religious, ritual, etc. However, their acquisitions are not always fully documented, and their archaeological or historical contexts are sometimes imprecise or poorly understood.

The dating of an object must enable it to be situated in the period of its creation, during which the artist or craftsman processed the material or reused it in a new work. This can then become a veritable chronological reference for other stylistically comparable works [DEF 08]. This is why art historians sometimes propose chronological estimates, based on iconographic, stylistic or technical elements, as well as archival documents.

3.6.2. *Choice of dating method(s)*

Archaeologists and anthropologists have long used a wide range of dating methods [BAH 19] to analyze the age of fossils, remains and other artifacts. However, this approach is somewhat rarer among those responsible for museum collections, whether archaeological, historical or otherwise. The main reason for this is undoubtedly the sampling that is often required on unique items [RIC 13a, RIE 21] or items that are sensitive from a deontological point of view, as in the case of ethnological objects [RIC 13b] or human mummies [RIC 17]. We can also imagine that heated controversies concerning the dates obtained for certain sacred [DAM 89] or world-renowned objects [DON 02, HIG 21] are not unrelated.

It is also true that so-called "laboratory" dating methods focus on the constituent materials of works or objects, and cannot intrinsically define the exact moment of their use, and therefore of their creation. Nevertheless, they are proving extremely useful in the study of museum works when the classical stylistic approaches of art history have reached their limits. They also make it possible to revisit or rediscover certain museum collections that have been neglected for lack of interest or documentation, and even to propose new chronologies for certain older periods [QUI 13].

Among dating methods, dendrochronology, luminescence and carbon-14 dating are the three most widely developed for museum works. Not all these methods concern the same materials or the same date ranges. In some cases, however, the combination of these three methods has enabled us to pinpoint the chronology of execution, as in the case of the wax portraits of Henri IV [LEH 11], made of several materials (wood, wax and terracotta).

Each of these methods has evolved considerably since their discovery. For carbon-14 dating, for example, the use of accelerator mass spectrometry since the 1980s has made it possible to reduce the size of samples by a factor of a thousand [STR 14], and sampling is now very minimally invasive (a few mg). Similarly, sample preparation protocols have improved and are becoming increasingly selective and targeted [DEV 18]. In the remainder of this chapter, we will present several selected examples of the contribution made by carbon-14 dating, a method that is undoubtedly the best known to the general public. Although simple in principle, it is tricky to implement, particularly when it comes to sample preparation and the often complex interpretation of results.

3.6.3. Authentication issues

The authentication of works of art is a real challenge, enabling us to refute or confirm an attribution. Many objects give rise to lively discussion and debate. The International Observatory on Illicit Traffic in Cultural Property advises that documentation should be gathered on art history, stylistic knowledge and technical or scientific characteristics, particularly dating. These are the three indispensable and necessary aspects for best practice in authentication and attribution.

An interesting example is the small icon known as "Notre-Dame de Grâce" in Cambrai Cathedral. This icon has been venerated since the 15th century, particularly during the August 15 procession, when it was carried through the streets of the city. Bequeathed to the cathedral in 1450 by Canon Fursy du Bruille, its origin is the subject of much debate. Periods of attribution range from the 10th to the 17th century, and even as far back as the 19th century, as numerous copies from this period have been identified [PHI 20]. The radiocarbon age obtained (820 ± 30 BP) corresponds to a date between 1175 and 1275, i.e. between the last quarter of the 12th century and the third quarter of the 13th century. This result, backed up by analysis of the paint layer, confirms the icon's age and Italian origin, and refutes speculation that it may have been copied in the 19th century.

3.6.4. Checking the validity of a date inscribed on the work

The first remark concerns the radiocarbon ages obtained for the wood samples taken from the Virgin's crown (210 ± 30 BP) and the child's arm (220 ± 30 BP). These two samples are therefore recent and correspond to restored pieces, as the restorer had correctly predicted. For the other samples, the radiocarbon ages are contemporary (wood and textile) and can therefore be combined. An average of the values is used to refine the result and the calculated statistic. The calibrated date obtained is then between 1265 and 1282, which is, to within 10 years, compatible with the date indicated on the base [TIC 21].

3.6.5. Tracing the history of a manuscript

A small manuscript consisting of 122 bound parchment pages was acquired by the Bibliothèque nationale de France. This unpublished manuscript (NAL 3245) retraces the life of St. Francis of Assisi on 16 pages, shedding new light on the history of the founder of the Franciscan order. This "rediscovered Franciscan manuscript" was studied in depth by a research group, whose findings led to major advances in Franciscan historiography [BUI 21].

A 12 mg parchment sample was taken from the lower part of f. 72, along with a fragment of the binding twine. The two calibrated date ranges obtained, between 1155 and 1265 for the binding twine and between 1215 and 1280 for the parchment, partially overlap. Therefore, the radiocarbon ages can be combined and the value obtained is between 1205 and 1270. Carbon-14 dating is compatible with the date predicted by Jacques Dalarun, who discovered the manuscript and proposed a date between 1230 and 1240.

The other major interest of this result lies in the dating of the binding twine, which is contemporary with the manuscript. It could in fact have corresponded to a second binding, probably carried out in the 15th century, as string was rarer before this date and visible holes on the spine of the manuscript indicated the presence of an old seam. The work was therefore bound twice in a relatively short space of time. One hypothesis is that the manuscript first benefited from a simple waiting binding before a definitive binding.

NOTE.–

– *Parchment* is the skin of an animal, usually sheep, sometimes goat or calf, specially prepared to serve as a writing surface. It is made up of numbered sheets. Example of a manuscript truncated at the beginning and end: three single sheets (f. 1-3) + two quires of eight sheets (f. 4–11, f. 20–27 and f. 28–35) + two single sheets (f. 36 and f. 43). The sheets can be assembled in different forms: the *volumen* is a set of sheets sewn end-to-end to form a scroll (used until the 4th–5th centuries) and the *codex* (used from the 1st–2nd centuries) is a set of sheets sewn into quires and can be considered the ancestor of the modern book;

4

General Information on Radiopharmaceuticals Used in Nuclear Medicine Imaging

Overall objective	
Provide a link between nuclear medicine and radiopharmaceuticals.	
Specific objectives	
Make the link between nuclear medicine and nuclear physics;	Know the principle of radiopharmaceutical administration;
Understand the objectives of nuclear medicine;	Understand the purpose of radiopharmaceutical quality control;
Make the link between radiotracers and radiopharmaceuticals;	Understand the usefulness of radiochemical purity;
Know the physicochemical criteria for choosing a radiotracer or radiopharmaceutical;	Know the methods for the experimental determination of radiochemical purity;
Knowing the types of diseases diagnosed in nuclear medicine;	Know the experimental methods used to determine radiochemical purity;
Define cancer;	Describe the experimental protocol of thin-layer chromatography applied to the determination of radiochemical purity;
Understand the origins of cancer;	Calculate radiochemical purity using thin-layer chromatography results;
Explain the mechanism of carcinogenesis;	Calculate radiochemical purity using the results of column chromatography;
Understand metastasis;	Describe the principle of high-performance liquid chromatography;
Define vasculogenesis;	Describe the principle of position emission tomography (PET);

For a color version of all figures in this chapter, see www.iste.co.uk/sakho2/nuclearphysics2.zip.

Understand the essential role of vascular endothelial growth factor (VEGF);	Describe the principle of single-photon emission computed tomography (SPECT);
Explain the mechanism of angiogenesis;	Know the nature of radioisotopes used in PET and how they are produced;
Describe tumor angiogenesis;	Know the nature and method of the production of radioisotopes used in SPECT;
Distinguish between angiogenic and inflammatory factors in cancer development;	Learn about the radioisotopes most commonly used in nuclear medicine imaging;
Differentiate between the incidence and prevalence of a disease;	Understand the usefulness and principle of PET scans;
Know how to calculate the incidence rate of a disease;	Understand the usefulness and principle of the PET/CT scan;
Define the concept of radiopharmaceutical products (RPPs);	Know the advantages and disadvantages of tomoscintigraphy;
Know the principle of the production of radioisotopes used in nuclear medicine;	Know the main scintigraphies and their uses.
Prerequisites	
Isotopia;	Mechanism of γ photon emission;
β+ emission mechanism;	Cell organization in the body.

4.1. Nuclear medicine

4.1.1. *Definition, objectives*

Nuclear medicine combines medicine and nuclear physics. As such, it encompasses all medical applications of radioactivity in medicine. Its two main aims are diagnostic (in vivo and in vitro) and therapeutic [MAR 10]. In both cases, a *labeled substance* is used, i.e. a substance into which a *radioactive isotope* of an appropriate element has been introduced. Once administered to a patient, the labeled substance becomes a *radiotracer* that travels to a biological tissue or organ that it selectively recognizes. The substrate (*organic* or *biological carrier*) to which the radioactive isotope is grafted is designed to promote a concentration of this isotope on the targeted tissue or organ. The radioactivity induced by the isotope is then either used to visualize its location (diagnosis), or to initiate damage to surrounding cells (therapy). The choice of a radioactive isotope (based on the nature of the radiation emitted, its physical properties (energy and half-life) and its chemical properties) will define the use of this molecule, known as a *radiopharmaceutical* [ZIM 06].

There are two major areas of application for nuclear medicine:

– *In vivo functional imaging*. This type of imaging involves administering a radiotracer to the patient, enabling it to be detected externally. In this field, we use scintigraphies based on the emission of gamma rays or tomographies based on the emission of positrons.

– *Radiation therapy*. In this field, X-rays or gamma photons are used to destroy cancer cells by breaking down their DNA. There are two types of this type of therapy: *External radiation therapy* and *internal radiation therapy*. External radiotherapy consists of directing ionizing radiation in high doses, 20–80 grays (Gy), depending on the tumor and organ, through the skin and tissues to destroy the tumor while sparing the surrounding healthy cells [INS 19]. *Metabolic radiotherapy* or *internal vectorized radiotherapy* treats benign (not serious) or malignant (serious) diseases. The radioactive isotope used is administered orally or by injection, and binds preferentially to the targeted diseased cells. There are two types of vectorized radiotherapy: *radioimmunotherapy,* in which radioisotopes are vectorized by *antibodies,* and *radiopeptidetherapy,* in which radioisotopes are vectorized by *peptides*. Brachytherapy is also another form of internal radiation therapy. The special case of brachytherapy of the prostate with an iodine-125 implant is discussed in section 3.7 of Chapter 3 of the third volume.

VOCABULARY CORNER.–

– *Antibody*: defensive substance generated by the body in the presence of an antigen whose toxic effect it neutralizes;

– *Antigen*: natural or synthetic macromolecule, which, when recognized by antibodies or cells of an organism's immune system, can trigger an immune reaction;

– *Peptides*: polymer of α-amino acids linked by peptide bonds.

4.1.2. *The birth of nuclear medicine*

Nuclear medicine originated with a major discovery in the traceability of radioactivity by Hungarian chemist George de Hevesy (1885–1966). This discovery is associated with the beginnings of radiodiagnostics. The various dates marking the birth of nuclear medicine are set out below.

1913: first use of radioisotopes as tracers in plant biology:

In 1913, Hevesy began work on radium D (^{210}Pb) at the Radium Institute in Vienna. Together with the British chemist Friedrich Paneth (1887–1958), who was working independently on the same subject, he developed a method for using radioisotopes as tracers (radiotracers) in chemical reactions. For two years, Hevesy tried in vain (he did not know the chemical nature of radium-D) to separate lead and radium-D by fine chemical methods. Radium D (RaD), so called because it was the fourth descendant of radium after RaA (Polonium 218, ^{218}Po), RaB (Lead 213, ^{213}Pb) and RaC (Bismuth 213, ^{213}Bi) [SAK 22] (Chapter 3, Table 3.7), is the radioactive isotope of lead (210 Pb) with a half-life of 22.26 years. This justified the failure of the separation: lead and radium-D are made up of the same chemical element. Therefore, having been unable to differentiate lead from "radium-D" by chemical methods, Hevesy thought that biological mechanisms would not allow this differentiation either. He therefore used radium-D as a "radioactive indicator" or tracer to study the properties of lead and explore its distribution in plants. Hevesy's experiment involved soaking the roots of a number of plants in a solution of lead nitrate and radium D. He then measured the radioactivity induced in the plants, measuring the radioactivity of the roots, then the stems and leaves. He then demonstrated that most of the lead passed into the roots of the plants used. Therefore, for a decade (1913–1923), Hevesy applied radioindicator-based analytical methods to various radiochemical studies, signing several publications with Paneth as co-author [MYE 79].

1923: first use of indicators in humans:

Indicators were first used in humans in 1923. The American physicist and cardiologist Herman Ludwig Blumgart (1895–1977) and the American physicist Soma Weiss (1898–1942) injected radium C (Bismuth-213) into an arm. They then measured the speed of blood flow between one arm and the other, as well as variations in this speed in patients with heart disease [ZIM 06]. However, the small number of natural radioelements available at the time hampered the development of the tracer monitoring method, particularly via IV (intravenous) injection.

1927: first arterial encephalography using sodium iodide injection:

In 1927, Portuguese neurosurgeon Egaz Moniz (1873–1955) performed the first arterial encephalography using sodium iodide injection [PAL 07, MAN 20].

1930: invention of the cyclotron for on-site production of very short-lived radiotracers:

The cyclotron was invented in the early 1930s by the American physicist Ernest Orlando Lawrence (1901–1958). It was the first particle accelerator and is used today in many large hospitals to produce in situ a number of very short-lived radioactive tracers. These include fluorine-18, used in the manufacture of radiopharmaceuticals such as 2-deoxy-[^{18}F]-2-fluoro-D-glucose, known by its acronym ^{18}FDG. The latter is the most widely used radiopharmaceutical in positron emission tomography (PET) [DEB 08].

1934: discovery of artificial radioactivity, paving the way for the production of various radioisotopes:

In 1933, French physical chemists Frédéric Joliot-Curie and Irène Joliot-Curie [SAK 22] (Chapter 3, Box 3.1) discovered artificial radioactivity [SAK 22] (Chapter 3, section 3.1). In 1933, while studying the effects of α radiation on matter, Frédéric and Irène Joliot-Curie observed that an inactive sheet of boron, aluminum or magnesium becomes radioactive when placed in front of a polonium source emitting α particles. When they moved the polonium source away, they noted that the radioactivity of the activated foil persisted, with β^+ positron emission. All the experimental results obtained were announced on January 15, 1934 at the Academy of Sciences.

1936: first injection of phosphorus-32 to treat leukemia:

In 1936, John Hundale Lawrence (1903–1991), an American physicist and physician and brother of Ernest Orlando Lawrence, injected a patient for the first time with radioactive phosphorus-32 to treat leukemia. John Lawrence is thus considered a pioneer in the field of nuclear medicine [ZIM 06].

1936: discovery of metastable technetium-99:

Technetium-99's metastable nuclide 99 m became the most important artificial isotope in imaging. It was discovered in 1937 by Italian-American physicist Emilio Gino Segrè (1905–1989), in association with Italian chemist Carlo Perrier (1886–1938) and American physicist Glenn Theodore Seaborg (1912–1999). In 1930, Segrè [SAK 22] (Chapter 3, section 3.6) discovered astatine and later plutonium-239, whose fissile nature he demonstrated. In 1935, Segrè drew up a map of nuclides, describing some of their properties graphically on a system of N/Z axes. Technetium-99 is still used in over 80% of nuclear medicine diagnoses.

Furthermore, the development of a technetium-99m generator contributed enormously to the boom in scintigraphy, by making the precious radioelement available to hospitals. Today, nuclear medicine is used in 100,000 hospitals worldwide. To meet demand, five reactors produce 95% of the Mo-99 required to manufacture ^{99m}Tc): the NRU (National Research Universal) reactor at Chalk River (Canada), HFR (High Flux Reactor) at Petten (Netherlands), BR2 (Belgian Reactor nr 2) at Mol (Belgium), OSIRIS at Saclay (France) and SAFARI (South Africa Fundamental Atomic Research Installation) at Pelindaba (South Africa). These reactors have been in operation for over 30 years. In the short term, some of them will be shut down, leading to a shortage of radioactive isotopes. If the production of these isotopes is not compensated for, the health repercussions could be significant [SOC 21].

1937: first use of sodium to treat thyroid disorders:

In 1937, the American medical physicist Joseph Gilbert Hamilton (1907–1957) first measured the thyroid uptake of iodine (iodine-128). The short half-life of this radioactive isotope (T = 25 minutes) limited its use. It was naturally replaced by iodine-131 (T = 8.02 days), which from 1938 onwards enabled comprehensive studies of thyroid metabolism and the diagnosis and treatment of hyperthyroidism (a thyroid disorder in which the thyroid gland produces an excess of thyroid hormones). These studies were carried out by Seaborg and the American physicist John Jacob Livingood (1903–1986).

Note that it was in 1937 that Livingood, British chemist Fred Fairbrother (b. 1921, date of death unknown) and Seaborg discovered the isotope 59 of the element iron. Iron-59 is a gamma and beta emitter with a half-life of 33.51 days. It is used in ferrokinetic tests to determine the rate at which iron is eliminated from plasma and incorporated into red blood cells.

1938: production of iodine-131 applied to the first thyroid studies:

In 1938, Livingood and Seaborg perfected the production of iodine-131 and cobalt-60. These three isotopes are still used today in nuclear medicine and radiotherapy. In 1938, American physicists Saul Hertz (1905–1950), Arthur Roberts (1912–2004) and Robley Dunglison Evans (1907–1995) carried out the first thyroid studies using iodine-131, treating the first patients with hyperthyroidism in 1932. Hertz is considered the pioneer in the use of radioactive iodine in nuclear medicine [HER 20].

In 1936, the American endocrinologist Samuel Seidlin (1895–1955), the radiophysicist Leonidas Marinelli (1906–1974) and Eleanor Oshry demonstrated that it was possible to eliminate all metastases in a thyroid cancer patient by treating with iodine-131. Then, the American physicist Benedict Cassen (1902–1972) demonstrated with radioactive iodine that thyroid nodules accumulate iodine, making it possible to differentiate benign from malignant cells. These results had a tremendous impact on the development of nuclear medicine, demonstrating beyond doubt the power of the method. For thyroid cancer treatment, this method remains indisputably the most effective. In 1951, iodine-131, in the form of sodium iodide, became the first radiopharmaceutical approved by the US Food and Drug Administration [ZIM 06, FEY 19].

1939: development of a technique for measuring blood volume using phosphorus-32:

In 1939, Hungarian chemist George Charles de Hevesy (1885–1986) developed a method for determining blood volume using phosphorus-labeled red blood cells. Hevesy used 1-milliliter injections of phosphorus-32-labeled red blood cells.

1950: origin of the term "nuclear medicine":

The term *"nuclear medicine"* was coined in the 1950s by American physician Marshall Brucer (1913–1993). He was a teacher and strong advocate of the use of radionuclides in medicine within the U.S. Army. He set up a course of study, enabling doctors to obtain the first authorizations to use radionuclides in the medical field.

1957: invention of the gamma camera and introduction of the first scintigraphies in the United States:

In 1957, American electrical engineer and biophysicist Hal Oscar Anger (1920–2005) invented the scintillation camera, known as the gamma camera (γ-camera) or Anger camera. Anger also developed a multi-planar tomographic radiation scanner using the Anger camera and a focused radiation collimator. Scintigraphy, by far the most widely used nuclear medicine examination, was the first to be developed after the appearance of Anger's gamma camera.

1963: first lung images using radiolabeled albumin aggregates:

In 1963, the American physicist Henri Wagner (1927–2012) produced the first lung images using radiolabeled albumin aggregates [FEY 19].

1971: invention of the scanner for image reconstruction:

In the 1970s, nuclear imaging took off, with the invention of the first CT scanner in 1971 by British engineer Geoffrey Hounsfield (1919–2004). The independent work of South African physicist Allan Cormack (1924–1998) and Hounsfield on the mathematical theory underlying scanner operation led to the development of computed tomography, also known as *CT scanning*. The invention of the CT scanner revolutionized morphological imaging, in particular radiology, thanks to decisive advances in nuclear medicine imaging. Since the advent of the scanner, the use of radiotracers has formed the basis of functional exploration, with associated techniques such as scintigraphy and PET, which may or may not be associated with a scanner (PET Scan). These techniques focus on the function (metabolism) of organs, tissues or cells, and are also precise diagnostic methods.

1973: development of the first PET camera:

The first *PET camera* was designed for human studies in 1973 at the University of Washington. It was invented by American nuclear chemist Edward Joseph Hoffman (1942–2004), medical physicist Michael Matthew Ter-Pogossian (1925–1996) and Michael Edward Phelps (b. 1939).

1975: development of positron emission tomography (PET):

PET was developed in 1975. As with scintigraphy, a decisive factor in the spread of PET was the development of a suitable radiopharmaceutical, fluorine-18 (FDG), at Brookhaven.

Every year, some 35 million people worldwide use nuclear medicine. This is either for diagnosis (90% of radiotracer use) or therapy (10% of radiotracer use). Global demand for radioactive isotopes is growing rapidly. The International Atomic Energy Agency (IAEA) estimates that several million new cases of cancer will be diagnosed each year worldwide, thanks to nuclear medicine. This represents real progress, since half of these diagnoses will be in developing countries [SOC 21].

4.1.3. *Diseases diagnosed in nuclear medicine*

In nuclear medicine examinations, the low-level radioactive tracers used accumulate specifically in certain tissues after injection. They can then be monitored from the outside by special cameras (*gamma-cameras*) to reconstruct a precise image of the activity in diseased organs and tissues. Scintigraphy and PET/CT scans enable doctors to estimate the severity of the disease, its extent and its progression. They can measure response to current treatments and, in the case of cancer, detect recurrence at an early stage. In this way, nuclear medicine enables therapy to be rapidly adapted to the evolution of the disease, leading to better treatment [CEN 20].

NOTE.–

A *PET scan* measures important bodily functions, such as metabolism. It helps doctors assess how organs and tissues are functioning.

Computed tomography (CT) imaging uses special X-ray equipment, and in some cases a contrast material, to produce several images of the inside of the body. A radiologist views and interprets these images on a computer screen. CT imaging provides excellent anatomical information.

Combined PET/CT scanners now perform almost all PET scans. These combined scans help identify abnormal metabolic activity and can provide more accurate diagnoses than the two scans performed separately [RAD 21].

In addition, nuclear medicine offers a non-invasive diagnostic tool (i.e. one that does not require the skin to be surgically broken or damaged). It is used to diagnose hundreds of pathologies, including cardiovascular diseases and cancers, as well as neurological disorders such as Alzheimer's and dementia (these diseases are discussed in Volume 3).

It should be noted that each type of disease has its own specific marker. For tumor examinations, radioactive sugar is absorbed by the tumor cells. This gives a very precise indication of their activity, enabling us to monitor the progress of the disease. In the case of cardiovascular disease, the marker binds to active areas of the heart and measures blood flow. For neurological disorders, the tracer indicates the activity of different brain areas by estimating the amount of oxygen absorbed [CEN 20]. In what follows, we look at the particular case of cancer.

4.2. Cancer

4.2.1. *Cell organization in the organism*

The human body is made up of approximately 60,000 billion cells with a predetermined life cycle [DES 13]. The body's different cell types divide identically as and when the body's needs change. These include growth (from infancy to adulthood), the renewal of hair, skin and blood cells, etc., and tissue repair through wound healing. Normal cells are thus born, multiply, die and renew themselves in a programmed way. As the life span of a cell is limited, this renewal takes place over generations. During these periods of regeneration, cell divisions are triggered by the body's need to repair damaged tissues. Therefore, at any given moment, the number of cells that die is balanced by the number of cells that are born.

Cell renewal is ensured by regulatory mechanisms. Each missing cell is replaced by a new cell with the same function and position in the body. Therefore, a normal cell is constantly subjected to molecules that dictate its transformation. Conversely, the normal cell in turn secretes molecules that influence the fate of neighboring cells [ZIM 06].

VOCABULARY CORNER.–

– *Cell*: the basic unit of life that makes up every organism, animal or plant. The human body is made up of several billion cells of different types (skin cells, bone cells, blood cells, etc.), most of which multiply, renew and die. Cells are visible under a microscope, and generally consist of a nucleus containing DNA and a cytoplasm bounded by a membrane. A cell becomes cancerous when it changes and multiplies uncontrollably.

– *Biological tissue*: an intermediate level of organization between the cell and the organ. A tissue is a collection of similar cells of the same origin, grouped together in a cluster, network or bundle (fiber). A tissue forms a functional whole, i.e. its cells work together to perform the same function. Tissues regenerate regularly and are assembled together to form organs (lung, heart, kidney, etc.).

– *Necrosis*: process of alteration leading to the death of a cell or tissue.

– *Cancer*: disease caused by the transformation of cells that become abnormal and proliferate excessively. These dysregulated cells eventually form a mass called a tumor.

– *Tumor*: a lump of varying size caused by an excessive multiplication of normal cells (benign tumor) or abnormal cells (malignant tumor). Benign tumors (e.g. moles, warts, etc.) develop in a localized area without affecting neighboring tissues. Malignant tumors (cancer) tend to invade neighboring tissues and migrate to other parts of the body, producing metastases.

– *Metastasis*: tumor formed from cancer cells that have broken away from an initial tumor (primary tumor) and migrated via lymphatic or blood vessels to another part of the body, where they have settled.

– *Lymphatic vessels*: tubes that circulate lymph throughout the body to the lymph nodes and back to the veins. The network of lymphatic vessels resembles the network of blood vessels (arteries and veins) that circulate blood.

– *Lymph*: whitish or yellowish biological fluid transported by the lymphatic system. Its composition is similar to that of blood plasma, of which it is merely a filtrate. It contains white blood cells, notably lymphocytes, and is devoid of red blood cells. It bathes the organs. Lymph is poorer in nutrients than blood, and richer in waste products. The average human body contains around 8–10 liters of lymph, compared with five liters of blood.

– *Red blood cells*: cells circulating in the blood and containing hemoglobin, which gives the blood its red color. Red blood cells transport oxygen.

– *Lymphocytes*: white blood cells or leukocytes that play an important role in the immune system (defending the body against infectious micro-organisms and foreign substances).

4.2.2. Evolution of cancer cells, tumor

The large number of cells dividing every day means that abnormal cells can appear either spontaneously or as a result of transformation induced by a carcinogen.

According to various studies, multiple factors can induce the development of abnormal cells in humans. These include smoking, diet, genetics, infections, occupational exposures, alcohol, obesity, UV exposure, drugs and pollution. These stressors act on cells, and are therefore possible causes of cancer. Indeed, in the face of these disturbances, certain genes may be deregulated, and a cell may undergo alterations or mutations of some kind [DES 13].

Naturally, such an altered or mutated cell can be recognized and destroyed by the immune system. However, abnormal cells sometimes survive because of their aggressive nature. These cells then multiply in an exaggerated and uncontrolled way, leading to the appearance of a tumor. The latter may be benign. In this case, it does not spread and poses no vital threat to the individual. A benign tumor can then be easily removed without the risk of recurrence. If the tumor is malignant, it invades healthy tissue to the point of inhibiting (or even destroying) its functions. Tumor cells are the origin of cancer.

The starting point of any cancer is therefore the cell which has been transformed and whose division has become uncontrollable because it escapes the organism's control. Thus, immortality, autonomous cell division and the ability to infiltrate are the main characteristics of a cancer cell [DES 13].

Regarding the actual causes of cancer, current research by various cancer organizations, such as the American Association for Cancer Research (AACR), shows that several factors directly related to people's lifestyles, such as smoking (one of the most determinant risk factors for stroke), physical inactivity, excess body weight, diet composition and immoderate use of alcohol and narcotics, are direct causes of the development of approximately 70% of cancers [BÉL 16].

4.2.3. *Carcinogenesis, metastasis*

By definition, *carcinogenesis* is the process of cancer formation. In the literature, it is also referred to as *oncogenesis*.

As mentioned above, a cancer is a collection of abnormal cells (tumor cells) that divide in a disorderly fashion. Malignant cells can break away from the original tumor and migrate via the bloodstream or lymphatic system to other parts of the body. In this way, cancer can invade various organs (breast, liver, lung, skin, testicle, etc.) or tissues of the body. The disease gradually spreads, leading to the appearance of *metastases* (new tumor colonies with properties identical to those of the original tumor cells) (Figure 4.1) [DES 13, BÉL 16].

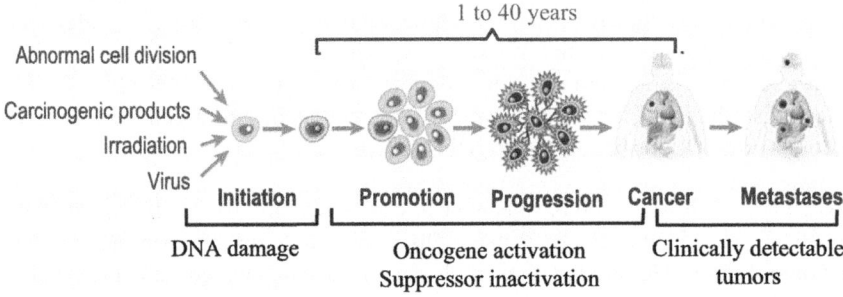

Figure 4.1. *Evolution of cancer. Image adapted from Béliveau and Gingras [BÉL 16]*

Each tumor cell can be identified by the organ from which it originates, as it is a malformation of a healthy cell of a specific type. For example, lung cancer that has metastasized to the liver is lung cancer that has progressed. Treatment will be for lung cancer, not liver cancer. Lymphomas and leukemias (also known as liquid cancers) involve the precursors of blood cells (hematopoietic system). These abnormal cells circulate in the blood and lymphatic systems, reproducing to the detriment of the production of healthy blood cells [ZIM 06].

NOTE.–

– Deoxyribonucleic acid, or DNA, is a biological macromolecule found in almost all cells as well as many viruses. DNA contains all the genetic information, called the genome, that allows the development, functioning and reproduction of living things. Denaturation of the DNA is the main cause of the onset of cancer (see Figure 4.1).

– Leukemias are cancers of the blood, which affect the cells of the bone marrow that give rise to white blood cells. Leukemia cells eventually invade the bone marrow and replace or suppress normal cell development.

– Lymphomas are characterized by the excessive proliferation of lymphocytes (usually B or T) in the lymph nodes, liver, spleen and more rarely other organs. They cause an increase in their size. The main symptoms are therefore an increase in the size of the lymph nodes. Diagnosis is based on biopsy analysis of the affected lymph nodes. Imaging examinations – CT scans and PET scans – will be carried out to assess the extent of the disease in the body. Since lymphocytes are blood cells, they can grow throughout the body: unlike solid tumors, the notion of metastasis does not exist in this kind of cancer. Source: https://curie.fr/dossier-pedagogique/cancers-du-sang-les-lymphomes, 2023.

– There are generally three groups of lymphocytes: B, T and NK (natural killer). Lymphocytes are synthesized within the bone marrow. They will then migrate to other lymphoid organs to continue maturing and can circulate in the bloodstream. Lymphocytes are white blood cells that play an important role in the immune system. NK lymphocytes, or NK cells, are involved in the innate immune response, which is the body's first response to an attack by pathogens. The innate immune response is immediate. The role of NK lymphocytes (also called killer cells) is to destroy altered cells such as infected cells and cancer cells. B and T lymphocytes are involved in the adaptive immune response. Unlike the innate immune response, this second phase of the immune response is called specific. Based on the recognition and memorization of pathogens, the adaptive immune response involves several leukocytes, including B lymphocytes that produce antibodies (complex proteins with the ability to recognize and neutralize pathogens in a specific way) and T lymphocytes that recognize and destroy pathogens in a specific way. Source: https://www.passeportsante.net›, 2023.

VOCABULARY CORNER.–

– *Oncogenes*: category of genes whose expression promotes the development of cancer. These are genes that control the synthesis of oncoproteins, proteins that stimulate cell division or inhibit programmed cell death. This triggers a disorderly proliferation of cells. The term *oncogene* can also refer to viruses that cause cancer. These are known as oncoviruses.

– *Suppressors*: a tumor suppressor gene or anti-oncogene is a negative regulator of cell proliferation. For example, TP53 is a tumor suppressor gene that controls cell growth and division. The TP53 gene also sends signals to other genes to help repair damaged DNA. If the damaged DNA cannot be repaired, the TP53 gene prevents the cell from dividing and tells it to die. When the TP53 gene is mutated, cells with damaged DNA begin to grow and divide in a disorderly fashion. Mutations in the TP53 gene are common, occurring in over 50% of all cancers (Changements génétiques et risque de cancer | Société canadienne du cancer, 2023).

Note that *TP53* stands for "tumor protein 53".

4.2.4. Angiogenesis, vascular endothelial growth factor (VEGF)

Vessels are created to supply tissues and cells with the oxygen and nutrients they need to function. The production of new *blood vessels* is essential for any developing tissue, including tumor tissue [COR 08, SÉG 15].

The cells of the human body need a constant supply of oxygen and nutrients to live. This supply is provided by the blood, which transports oxygen from the lungs to the cells, and nutrients from the intestine to the cells. To be served by blood circulation, cells need to be fairly close to blood vessels. For this reason, the *vascular network* is highly developed to ensure the proper supply of oxygen and nutrients to all the body's cells. Toxic waste produced by cells is carried by the bloodstream, filtered by the kidneys and eliminated from the body through urine.

Angiogenesis processes are generally extinguished during adult life, due to the stability of the vascular network. Vasculogenesis is the very first stage in the formation of blood vessels. Vascular endothelial growth factor (VEGF) is a biological factor that plays an essential role in normal and pathological vasculogenesis and angiogenesis.

When tissue is growing, mainly during development or wound healing, the new cells generated gradually move away from the blood vessels as the tissue increases in thickness and size (Figure 4.2).

To ensure that new tissue cells continue to be properly served by the bloodstream, new blood vessels form from nearby ones. This is made possible by the proliferation of the cells that make up blood vessels: endothelial cells. This physiological process is called *angiogenesis*. The new blood vessels grow in the direction of a tissue area poorly supplied by the bloodstream. Indeed, tissues that receive too little oxygen and nutrients generate angiogenic signals such as VEGF that stimulate angiogenesis, thus vascularizing this undernourished area and enabling it to be served by the bloodstream. Once the tissue is properly vascularized, angiogenesis ceases [SÉG 15], as shown in Figure 4.2.

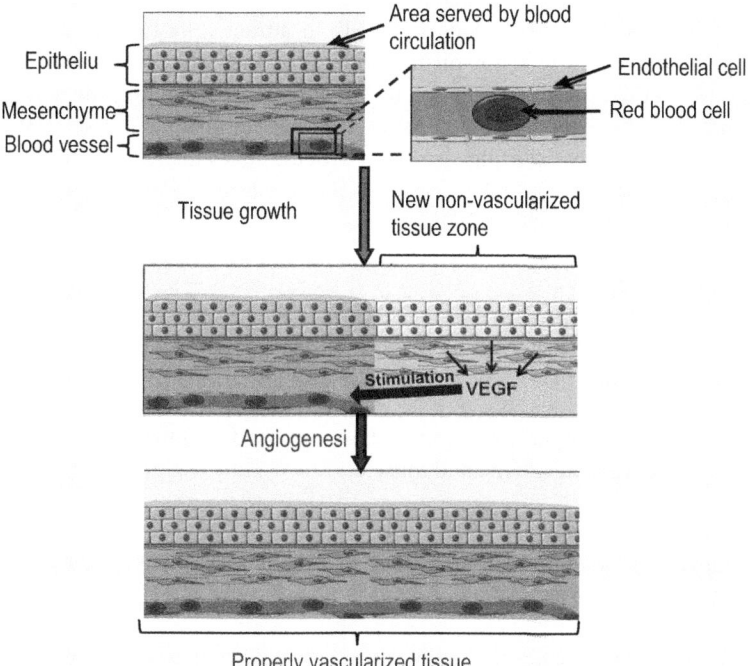

Figure 4.2. *Angiogenesis. Tissue growth gives rise to a new area of tissue that is poorly served by the bloodstream. Cells in this non-vascularized area secrete VEGF, which stimulates the proliferation of endothelial cells. This stimulation causes new blood vessels to sprout in the direction of this area, enabling it to become vascularized. Image adapted from Ségala [SÉG 15]*

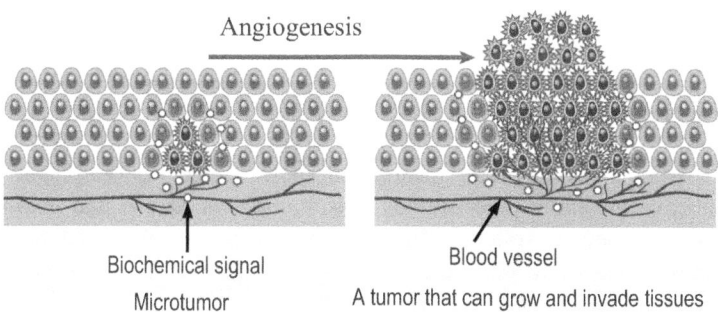

Figure 4.3. *Angiogenesis, an essential process in tumor growth. Image adapted from Béliveau and Gingras [BÉL 16]*

In tumor angiogenesis (see the next section), cancer cells produce chemical signals (angiogenic signals), including VEGF, to attract cells from a nearby blood vessel. By binding to a receptor on the surface of the cells in the vessel, VEGF induces these cells to make their way into the tumor by dissolving the surrounding tissue, and to form enough new cells to make a new blood vessel [BÉL 16]. Therefore, angiogenesis, as a physiological process that enables the production of new blood vessels, is essential to any developing tissue (Figure 4.2), including tumor tissue [SÉG 15], as shown in Figure 4.3.

VOCABULARY CORNER.–

– *Epithelium*: tissue made up of juxtaposed cells that covers the surface of the body or lines the inside of all hollow organs.

– *Mesenchyme*: embryonic connective tissue (mesoderm) that gives rise to adult connective tissue, cartilage, bone and muscle.

– *Connective tissues*: these support the body's other tissues, providing nutrition and participating in the body's immune defense mechanisms. They are disseminated within and between organs.

– *Endothelial cells*: cells that play an active and essential role in controlling vascular tone, and therefore local blood flow.

4.2.5. *Tumor angiogenesis*

A primary tumor can develop in various organs (breast, liver, lung, skin, testicle, etc.). Clusters of cancerous cells can then proliferate by seeking nutrients in their immediate environment via the formation of new blood vessels. This constitutes *tumor angiogenesis* (from the Greek *angio*, vessel, and *genesis*, formation) (Figure 4.2). Sometimes, cancer can be detected even before it has metastasized. In such cases, it is important that treatment can be applied as quickly as possible to limit tumor progression [DES 13].

Two particularly important factors promote cancer development in the immediate environment of precancerous cells: *angiogenic factors* and *inflammatory factors* of the immune system. The concerted action of these factors enables precancerous cells to draw on the elements necessary for their progression from their immediate environment.

– Angiogenic factors are determined by angiogenesis which, as shown in Figure 4.3, contributes to tumor progression by providing a new network of blood vessels (Figure 4.2). This new network enables the tumor, on the one hand, to meet its energy needs and, on the other hand, to continue invading surrounding tissues (Figure 4.3).

– Inflammatory factors are determined by the inflammation caused by our immune system. This inflammation is a biological phenomenon essential to the entire human body. Without it, the organism would be entirely at the mercy of the many pathogenic agents present in our environment (see the note below). However, when inflammation becomes too intense or occurs over too long a period, it can lead to the development of a number of pathologies, and even promote the progression of diseases such as cancer [BÉL 16].

NOTE.– Inflammation, an ally that can also become an enemy.

The immune system is the set of phenomena that enable us to defend ourselves against aggression, whether pathogenic (bacteria, viruses), chemical or traumatic in origin. This system is a veritable armed force made up of elite soldiers divided into groups specialized in specific tasks of neutralization or attack. The "inflammatory squad", the division in charge of rapidly neutralizing intruders, is on the front line. The cells in this squad, in particular certain white blood cells called macrophages, are known as "inflammatory" because they release highly reactive molecules designed to eliminate any pathogens attempting to invade our body. This causes irritation (easily spotted in the form of redness, swelling or tingling). This inflammatory reaction also serves to initiate the repair of damaged tissue, thanks to the many growth factors secreted by inflammatory cells. These accelerate the arrival of healthy cells and promote the formation of new blood vessels. Under normal circumstances, this reaction should be short-lived, as the continuous presence of inflammatory molecules becomes extremely irritating for the tissues involved. When it persists, a state of chronic inflammation sets in, which can lead to intense pain at the site of inflammation. As we shall see, chronic inflammation can also be favoured by certain lifestyle factors (smoking, obesity, caloric overload, omega-3 fatty acid deficiency). Although this type of chronic inflammation does not necessarily cause any apparent symptoms, it nevertheless creates a climate conducive to the growth of cells present in the inflamed environment. This state is particularly dangerous if the tissue contains microtumors composed of precancerous cells. These can then use the growth factors secreted by the inflammatory cells and the new network of blood vessels created in the vicinity of the inflammation to become a mature tumor [BÉL 16].

4.2.6. *Global cancer epidemiology data*

In this section, we summarize global data on cancer epidemiology from the Association de Recherche sur les CAncers dont GYnécologiques (Association for Research into Cancers including Gynecological Cancers) [ASS 20].

– Worldwide in 2018:

According to data from the IARC (International Agency for Research on Cancer), it is estimated that in 2020, the worldwide incidence (new cases diagnosed) of cancer would be approximately 18.6 (18–19.3) million. Its worldwide prevalence (people affected and alive) at five years would be approximately 33 million. This disease is the second leading cause of death after cerebrovascular disease. In 2015, it accounted for 8.8 million deaths, or 17% of all deaths. Table 4.1 shows cancer locations by incidence and prevalence, as well as induced mortality by affected organ.

Most frequent locations in incidence	Most frequent locations in prevalence	Mortality by organ affected
Lung: 12.7%	Breast: 17.7%	Lung: 21%
Breast: 10.9%	Colon and rectum: 10.6%	Liver: 9%
Colon and rectum: 9.3%	Prostate: 6.9%	Colon and rectum: 8.8 %

Table 4.1. *Most frequent locations according to cancer incidence and prevalence, and induced mortality according to the organ affected*

The five-year survival rate, all cancer sites included, is 67% in industrialized countries. The standardized mortality rate per 100,000 is estimated at 131.8 for men and 77.5 for women, corresponding to an M/F ratio of 1.82.

– In the European Union:

Globally, according to WHO statistics, the number of incident cases of cancer is approximately 3.9 million (2.05 million M; 1.85 million F), excluding skin cancers but including melanoma. The estimated number of cancer deaths in 2018 was nearly 1.93 million (M: 1.08 million; F: 850,000).

In 2020, the EU accounted for 55% of the European population. By 2020, there would be 2.7 million new cases (all types combined, excluding non-melanoma skin cancer) and 1.3 million deaths.

In France, the overall cancer epidemiology data from the Institut National du Cancer (National Cancer Institute) [INS 21] are as follows.

The estimated number of new cancer cases in 2018 was 382,000 (53% in men, 36% in women).

An estimated 382,000 new cases of cancer were diagnosed in 2018 in mainland France (203,600 men and 177,300 women). Incidence rates (world standardized) were estimated at 330.2 per 100,000 men and 273.0 per 100,000 women. The median age at diagnosis was 68 for men and 67 for women. The number of cancer deaths was estimated at 157,300 in 2018: 89,600 in men and 67,800 in women (vs. 83,031 men and 66,000 women in 2017). Estimated mortality rates (world standardized) were 123.8 per 100,000 men and 72.2 per 100,000 women. The median age at death was 73 for men and 77 for women. In men, prostate cancer remains by far the most common (50,300 new cases in 2015), ahead of lung cancer (31,200 cases in 2018) and colorectal cancer (23,000 cases in 2018). In women, breast cancer leads the way (58,500 cases in 2018), ahead of colorectal cancer (20,100 cases) and lung cancer (15,100 cases).

Between 2010 and 2018, the number of new cancer cases rose by 6,060 in men and 23,053 in women. However, the *standardized incidence rate* fell in men (-1.3%) and is tending to stabilize in women (+0.7%). In terms of mortality, lung cancer ranks first in men (22,800 deaths in 2018), ahead of colorectal cancer (9,200 deaths) and prostate cancer (8,100 deaths).

In women, breast cancer is the leading cause of cancer deaths (12,100 deaths in 2018), ahead of lung cancer (10,300 deaths) and colorectal cancer (7,900 deaths). Over the same period (2010–2018), the *standardized mortality rate* (SMR) fell by 2% a year for men and 0.7% a year for women. The overall reduction in mortality is the result of earlier diagnosis and major therapeutic advances, particularly among the most common cancers.

In 2018, some 3.8 million people aged 15 years and over had cancer in their lifetime. On average, their risk of a second cancer was 36% higher than that of the general population.

In March 2023, researchers were able to predict that 1,261,990 people would die of cancer in the European Union (EU-27) in 2023 and 172,313 in the UK [MAL 23].

– In sub-Saharan Africa:

Sub-Saharan Africa is made up of 48 countries, including Senegal, Gambia, Côte d'Ivoire, and South Africa (see: https://www.mcgill.ca/mastercardfdn-scholars/liste_des_pays_admissibles_2020.pdf).

The figures for cancer in this region of sub-Saharan Africa in 2020 are as follows [ANK 22].

The new report provides an overview of cancer trends in terms of numbers and characteristics of cancers in the region. It shows an estimated total of 801,392 new cancer cases and 520,158 cancer deaths in sub-Saharan Africa in 2020. Female breast cancer (129,300 cases) and cervical cancer (110,300 cases) were responsible for three out of every 10 cancers diagnosed in both sexes. Among women, the most common types of cancer are breast cancer (in first place in 28 countries) and cervical cancer (in 19 countries). For men, the most common type of cancer was prostate cancer (77,300 cases), followed by liver cancer (23,700 cases) and colorectal cancer (23,300 cases). Prostate cancer is the leading incident cancer in men in 30 sub-Saharan African countries. The risk of a woman in sub-Saharan Africa developing cancer by the age of 75 is 13.1%, with breast cancer (3.1%) and cervical cancer (3.5%) together accounting for half of this risk.

NOTE.–

Vitamin D acts like a steroid hormone via receptors distributed throughout the body, and not just in the intestine, to promote active calcium absorption. This implies that this vitamin has extraosseous effects on the immune system and the regulation of cell proliferation. However, it is important to note that, theoretically, vitamin D should not be considered as a vitamin *"defined as a vital product that the body cannot produce"*, since its synthesis is carried out in the skin under the effect of insolation (under the action of UVB rays of particular wavelength: 290–315 nm) [MAL 13].

NOTE.–

Ultraviolet rays, known as UVA and UVB, have different effects on the skin:

– UVA rays have a wavelength of 300–315 nm. They penetrate deeper into the dermis, and are mainly responsible for pigmentation spots, skin aging and wrinkles. They can also promote the appearance of skin cancers.

– UVB rays have a wavelength of 315–280 nm. They penetrate less deeply and are responsible for sunburn (red, painful patches caused by UV-induced burning), burns, blisters and most skin cancers.

The effects of the sun can also be long term. It can take up to 20 years from the onset of sunburn to the appearance of skin cancer [MAL 13].

Figure 4.4 shows variations in the incidence of a disease as a function of time. These variations correspond to the evolution of the disease. Incidence is stable during the endemic phase, rises during epidemics and pandemics, or falls to zero when no new cases are reported.

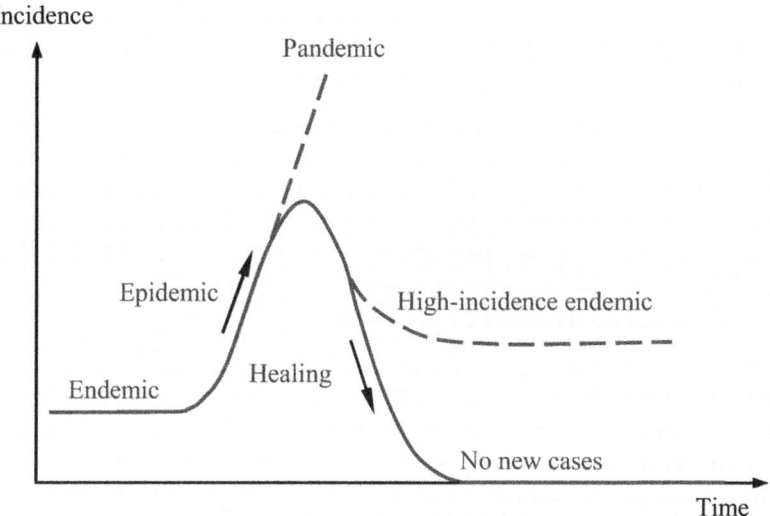

Figure 4.4. *Schematic representation of possible changes in the incidence of a disease over time*

VOCABULARY CORNER.–

– *Melanoma*: a potentially aggressive tumor, especially when caught at a late stage, which can give rise to life-threatening metastases. 65–95% of cutaneous melanomas are caused by exposure to UV radiation.

– *Epidemiology*: scientific discipline concerned with the study of health problems in human populations, their frequency, distribution in time and space, and the factors influencing population health and disease.

– *Incidence*: in epidemiology, the incidence of a disease is the number of new cases over a given period.

– *Incidence rate*: the incidence rate is calculated by dividing the number of new cases (incident population) by the size of the target population, over a given period of time. This rate is one of the most important criteria for assessing the frequency and rate at which new cases of a disease appear.

– *Prevalence*: a measure of the proportion of the population affected by the disease at a given time.

– *Endemic*: refers to the habitual presence of a disease in a given region. The term endemic is used when an infectious and contagious disease becomes permanently established in a given region. The disease is either permanent or latent, affecting a significant proportion of the population. The disease is known and identified, but this does not mean that it is progressing or spreading. Yellow fever and malaria are well-known examples of endemic diseases.

– *Epidemic*: the rapid development and spread of a contagious disease, usually of infectious origin, among a large number of people. It is therefore limited to a well-defined region, country or area. For example, Ebola is one of the world's most deadly viruses. Transmission occurs through direct contact with body fluids. Between 2013 and 2015, an Ebola epidemic caused the death of 11,323 people in West Africa. See: https://www.cnews.fr/monde/2020-05-27/voici-les-7-virus-les-plus-dangereux-de-la-planete-782382.

– *Pandemic*: an epidemic that affects a large number of people over a very wide geographical area. This is the case with Covid-19 caused by SARS-CoV-2, or the "new coronavirus". The main symptoms of this highly contagious virus are fever, dry cough and fatigue. As of January 8, 2023, the number of coronavirus-infected people worldwide, by country, was 657.97 million. See: https://fr.statista.com/statistiques/1091585/morts-infections-coronavirus-monde/.

NOTE.– Calculating the incidence rate.

In epidemiology, the incidence I in a population is expressed as the number of new cases per person-time (i.e. the cumulative follow-up time of at-risk individuals in the population studied). By definition:

$$I = \frac{N}{P \times D} = \frac{N}{T} \qquad\qquad [4.1]$$

In relation [4.1], N is the number of new cases, P is the total number of individuals in the study population, D is the length of the observation period, and T is the total duration expressed in person-years.

As an illustrative example, if 100 people at risk ($P=100$) were studied for three years ($D= 3$), the total duration of follow-up is $T = P \times D = 100 \times 3 = 300$ person-years. If there were 15 new cases of the disease observed during the study

($N = 15$), then the incidence rate $I = 15/\ 300 = 0.05 = 50\ \%$ (15 cases per 300 person-years, or five cases per 100 person-years). Note that the incidence in a population must be "at risk", i.e. members of the population must be able to contract the disease.

4.2.7. Cancer control in Senegal

The IAEA and its partners, the World Health Organization (WHO) and the International Agency for Research on Cancer (IARC), carried out the review from July to December 2020 in virtual form due to travel restrictions caused by the coronavirus. The review shows that Senegal could have one of the most advanced cancer control systems in West Africa, says Geraldine Arias de Goebl, Head of the Cancer Control Review and Planning Section at the IAEA.

The review was carried out under the direction of the Ministry of Health, with the participation of civil society organizations and ten public and private healthcare centers from all over the country. All stages of cancer control were examined: planning, surveillance, prevention, early detection, diagnosis, treatment and palliative care. Discussions also covered radiological safety and security considerations. The review resulted in specific recommendations for certain cancers, such as pediatric cancer and cervical cancer.

In 2020, IARC estimated that more than 11,000 people were diagnosed with cancer in Senegal each year, and that nearly 8,000 died from it. *Cervical cancer* is the leading cause of cancer mortality in the country, and one of the five leading causes of death in general. These figures are expected to increase substantially, reaching over 16,000 new cases and more than 11,000 deaths by 2030 [AGE 21].

In addition, in an article published on February 2023, the Secretary General of the Ministry of Health and Social Action of Senegal pointed out that "cervical, breast, liver, prostate and stomach, account for nearly half of the new cancer cases in the country". An evil that does not spare the little ones either. "Children are also affected and each year, the paediatric oncology unit receives an average of 220 new cases out of the expected 800" [NDA 23].

Prostate cancer is the most common cancer in men. Prognosis is generally good, with a survival rate of over 90%. It remains rare before the age of 50, and the number of cases increases progressively after the age of 50 [SCH 21]. In our recent work [NGU 23], we studied the risk of secondary cancer in the particular case of the prostate for 30 patients aged between 52 and 81 years, an age group in which the number of cases of prostate cancer continues to rise.

4.2.8. Recommendations from cancer organizations

We summarize the recommendations of the cancer control organizations in Table 4.2 [BÉL 16].

	Risk factors		Recommendations
	Smoking		Quitting smoking
Carcinogenic agents	Excessive drinking		Limiting consumption to two glasses a day for men and one glass a day for women.
	Excessive exposure to UV rays		Protecting your skin from the sun by avoiding unnecessary exposure. Avoiding exposure to artificial sources of UV rays (tanning booths).
	Sedentary lifestyle		Being physically active for at least 30 minutes a day
	Plant deficiency		Eating a wide variety of fruits, vegetables, legumes and whole grains.
	Overweight and obesity		Staying as slim as possible, with a body mass index between 21 and 23.
Diet and body weight control	Industrial food (junk food)		Avoiding fizzy drinks and minimizing consumption of high-energy foods containing high amounts of sugar and fat.
	Excessive red meat and deli meats		Reducing consumption of red meats (beef, lamb, pork) to around 500 g per week, replacing them with fish, egg or vegetable protein-based meals. Keeping deli meats to a minimum.

	Excessive salt		Limiting consumption of products preserved in salt (e.g. salted fish) and products containing a lot of salt.
	Consumption of supplements		Not compensating for a poor diet by using supplements: the synergy offered by a combination of foods is far superior in reducing the risk of cancer.

Table 4.2. *Recommendations from cancer organizations*

NOTE.–

The body mass index (BMI) is used to assess your body size and adapt your lifestyle, if necessary, to avoid becoming overweight or thin. BMI has been used since 1997 by the World Health Organization (WHO), primarily to assess the risks of being overweight and obesity in adults aged 18–65 years. This indicator allows us to know if our weight is adapted in proportion to our height or if it presents a potential danger to our health. BMI is calculated by dividing mass (expressed in kilograms) by height squared (expressed in meters), i.e.: BMI = body mass (in kg)/height^2 (in meters). The WHO defines several alert thresholds:

– below 18.4 kg/m^2, the person is considered to be thin;

– between 18.5 and 24.9 kg/m^2, the person is considered to have a "normal" build;

– between 25 and 29.9 kg/m^2, the person is considered to be overweight;

– between 30 and 34.9 kg/m^2, the person is considered moderately obese (1st stage of obesity);

– between 35 and 39.9 kg/m^2, the person is considered to be severely obese;

– above 40 kg/m^2, the person is considered morbidly obese [DUR 23].

It should be noted that obesity is considered morbid when the risk of one or more serious health problems or diseases resulting from obesity increases considerably. These comorbidities can cause severe disability or even death.

4.3. General information on radiopharmaceuticals

4.3.1. *Notion of radiopharmaceuticals, specific properties*

Nuclear medicine uses the properties of certain artificial radioactive isotopes, which may or may not be associated with a carrier molecule, to diagnose, monitor and manage a large and varied number of pathologies, as well as to treat some of them with *metabolic radiotherapy*. The radioactive isotope is called the *marker*, and the molecule with which it may be associated is called the vector. Together, they form the radiopharmaceutical or radiotracer. The radiotracer acts as a microemitter of gamma rays, which can be detected using devices such as gamma cameras [OPE 07]. Once injected into the bloodstream during various imaging tests (PET, scintigraphy, etc.), the tracer can be visualized in the patient's body. By binding to different organs, the tracer can be used to analyze organ function [INS 10].

Radiotracers are defined as radiopharmaceuticals (RPPs). A drug is defined as any substance or composition presented as having curative or preventive properties with regards to human or animal diseases. It is also defined as any product that can be administered to humans or animals, with a view to establishing a medical diagnosis or to restoring, correcting or modifying organic functions [PAY 08]. In addition, a RPP, the concept of which first appeared in 1965, is defined as "a drug whose active principle is based on the properties of the radioactive emission of a radioelement" [BOU 17, MAN 20]. RPPs are generally used for diagnostic purposes (85% of cases), and more rarely for therapeutic purposes (15% of cases) [THO 18].

In addition, radioactive labeling can be carried out in two ways; this is either by replacing a stable atom in a molecule with one of its radioactive isotopes, or by attaching an additional, radioactive atom to a molecule. The radiotracer is chosen on the basis of its radioactive half-life, which must be sufficiently short for the tracer mass to be very low, but still correspond to a detectable activity. It is also chosen for the nature and energy of the radiation emitted [CEA 18].

Unlike morphological imaging, where the radiation is external to the body, the radiation used in nuclear medicine imaging is internal. The biological support molecule is used to monitor the activity of specific organic functions in living tissues. The isotope added must not alter the biological function of the pharmaceutical molecule, while maintaining a strong bond with it. The energy of the photons emitted must be in line with the patient's absorption and the system's detection efficiency, as well as having a radioactive half-life suitable for chemistry, medical examinations and dosimetry [BEK 06, COU 18].

The radioisotopes used in nuclear medicine are all artificially produced. They are radioelements produced either in a nuclear reactor or in a particle accelerator (cyclotron). As for the carrier molecules used in the manufacture of RPPs, these are marketed in kit form. The RPP synthesis process can be translated formally by the equation:

Vector (kit) + Marker (radioisotope) = Specific radiopharmaceutical drug.

From a practical point of view, radiopharmaceuticals are administered to the patient by intravenous injection, ingestion, inhalation or other means. They may be isolated radionuclides, such as iodine-123 for the thyroid gland, or consist of a carrier and a radionuclide. A carrier molecule, part of the human metabolism, is attracted to a target organ. The resulting radiotracer emits ionizing radiation that can be detected and used, for example, to destroy cancer cells.

4.3.2. *Quality control of radiopharmaceuticals*

As explained above, radiopharmaceuticals are RPPs in the form of sterile, pyrogen-free (does not cause or increase fever) solutions, which can be injected into patients for diagnostic or therapeutic purposes. In terms of purity and efficacy requirements, RPPs are no different from conventional injectable medicines (CIMs), such as tetanus vaccines. However, there is a fundamental difference between RPPs and CIMs in terms of their lifetime. While CIMs have an expiry time expressed in months or years, due to their radioactive nature, RPPs have a limited lifetime expressed in days or hours. For this reason, RPPs have to be manufactured, tested and administered to the patient in a very short space of time, usually on the same day.

In general, the aim of quality control for radiopharmaceuticals is to ensure that the finished marketable product meets the quality requirements set out in its marketing authorization. Within the framework of the European Pharmacopoeia, radiopharmaceutical controls are carried out in accordance with the tests described in the monograph "Radiopharmaceutical preparations". These tests include [DES 00]:

– physico-chemical controls: pH, identity, color, clarity, active ingredient content, radiochemical purity (RCP);

– radiometrological controls: radionuclidic purity and radioactive concentration;

– biological controls: sterility, bacterial endotoxins, absence of abnormal toxicity.

By definition [DES 00, BOU 17]:

– *Radiochemical purity* is the ratio of the radioactivity of the isotope in question, present in the product in the chemical form indicated, to the total radioactivity of the same radioactive isotope present in the product.

– *Radionuclidic purity* is the ratio of the radioactivity of the isotope in question to the total radioactivity of the radiopharmaceutical.

– The abnormal toxicity test is designed to verify the safety of the product by injection into the animal: no animal should die during the seven-day observation period following injection of the product.

NOTE.–

– Radionuclide purity is determined by γ spectrometry and monitored with an activimeter (a particle detector used to measure the activity of a radioactive source). The control of radionuclide purity makes it possible to highlight isotope errors, the presence of impurities which would result in abnormally high irradiation of the patient and related poor quality images to the characteristics of the impurities. To do this, a few microliters of final solution are placed in front of a calibrated gamma spectrophotometer (NaI). A single peak should be observed at 511 keV [MAJ 14].

– Endotoxins are complex toxic molecules, components of the wall of certain bacteria.

– The importance of pH control is justified by the fact that it conditions the complexation of the radionuclide by the vector used. The pH value must be compatible with that of the physiological pH (5.5–8), and must lie within the pH range for which the preparation is most stable. pH control can be easily performed using pH paper: place a drop of the preparation on a pH paper and compare it with a colorimetric scale (written on the roll of pH paper).

4.3.3. *Radiochemical purity, experimental determination methods*

Over 90% of RPPs are prepared in the Nuclear Medicine laboratory before injection into the patient. In this case, the physician is provided with a pure radionuclide and a non-radioactive chemical compound packaged by the pharmaceutical industry for labeling with the radionuclide. The important parameter for the physician is the labeling yield, which they check themselves just before injecting the patient [WAS 98].

In RPP synthesis, the most important parameter affecting diagnostic efficiency is radiochemical purity (RCP) [WAS 98, ROU 11, BOU 17]. In the body, radiochemical impurities have a different biodistribution. This results in disruption of the scintigraphic image, with the consequent risk of diagnostic error and unnecessary irradiation during the examinations to be carried out. To check the quality of the radiopharmaceutical formulation, the RCP is determined for each product or each production batch.

Three experimental methods can be used to determine RCP: thin-layer chromatography (TLC), column chromatography and high-performance liquid chromatography (HPLC), a highly accurate but time-consuming method that is rarely used on a routine basis. The principles of the first two techniques are outlined below.

– *Thin-layer chromatography (TLC)*:

TLC is *based on electrostatic interaction (H-bonding)*. Its principle consists of separating the constituents of a mixture according to their affinity to a stationary phase fixed on a plate, and a mobile phase called the eluent (solvent or mixture of solvents) that passes through (diffuses) the stationary phase. Molecules with a high affinity for the stationary phase will migrate more slowly than those with no affinity. For this technique, RCP is determined by the equation:

$$\text{RCP (\%)} = [\text{RPP activity} / (\text{RPP} + \text{impurities}) \text{ activities}] \times 100 \qquad [4.2a]$$

– *Column chromatography*:

This technique is similar to TLC, except that elution is by gravity rather than diffusion. A glass column equipped with a sintered glass and a tap is used. The column is filled with a powder, usually alumina or silica. A mixture is placed at the top of the column. Depending on the nature of the eluent and the contents of the column, some molecules are more easily eluted than others. For this technique, RCP is determined by the equation:

$$\text{RCP (\%)} = [A/(A + B)] \times 100 \qquad [4.2b]$$

In equation [4.2b], A is the column activity and B is the eluate activity.

The eluates and column are collected in tubes and counted using an activimeter.

In conclusion, RCP must be greater than or equal to 95%. This percentage means that for ^{18}FDG, for example, the majority of fluorine-18 (over 95%) must be in the chemical form ^{18}FDG [BOU 17]. If RCP is below this specification (95%), the preparation must be discarded [AGE 13b].

– High-performance liquid chromatography (HPLC):

The compounds to be separated (solutes) are dissolved in a solvent. This mixture is introduced into the liquid mobile phase (eluent). Depending on the nature of the molecules, they interact to a greater or lesser extent with the stationary phase in a tube known as a chromatographic column. The mobile phase is pushed through the chromatographic system by a high-pressure pump. The mixture to be analyzed is injected and transported through the chromatographic system. The compounds in solution are then distributed according to their affinity between the mobile phase and the stationary phase. At the column outlet, the various solutes are characterized by a peak using an appropriate detector. The set of recorded peaks is called a chromatogram [SAK 96, SAK 97]. We applied this high-performance method in our project for the analysis of drinking water in three regions of Senegal, using a conductimetric detector at the outlet of the column [SAK 96, SAK 97]. The principle of TLC is discussed in detail in the following section.

4.3.4. *Thin-layer chromatography applied to the determination of radiochemical purity*

TLC is an excellent method for separating and identifying the constituents of a homogeneous mixture (liquid or gas). Chromatography is based on differences in behavior between a common mobile phase and a stationary phase. In the case of gas chromatography, the mobile phase is a carrier gas. In paper, thin-layer or column chromatography, the mobile phase is a liquid called the eluent.

The stationary phase may be solid or liquid. Solids such as silica (SiO_2) or alumina (Al_2O_3), after treatment, can be used to separate components from mixtures, thanks to their adsorbent properties. These solids can be used as column packing in the case of gravity chromatography or high-performance liquid chromatography (HPLC) [WAS 98, ROU 11]. They can also be spread in a thin layer on a glass, aluminum or plastic sheet, as in the case of TLC. Gravity chromatography uses silica particles between 70 and 200 μm in size, and the solvent is dripped off. This technique is obsolete, as it requires larger quantities of silica and solvent.

In line with the objectives set out in this chapter, chromatography can be used to experimentally determine the RCP of a given radioactive product. For this reason, we briefly present its principle in what follows.

In TLC, the constituents of a homogeneous mixture are separated by transferring a solvent onto a support. The solvent is often referred to as the eluent or mobile phase, and the support as the stationary phase. The experimental set-up for TLC is shown in Figure 4.5(a) [MAG 21].

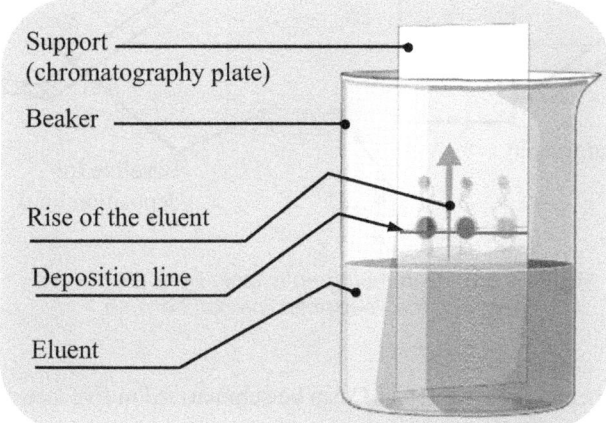

Support
(chromatography plate)

Beaker

Rise of the eluent

Deposition line

Eluent

Figure 4.5(a). *Experimental set-up for thin-layer chromatography. Note the deposition line used to determine radiochemical purity*

The principle of TLC is as follows:

1) Deposit a small quantity of the mixture to be separated on the support.

2) Support in contact with the solvent or eluent used.

3) Upward migration of the eluent along the support by capillary action.

4) The *elution phenomenon* begins: the eluent carries the constituents of the mixture to the top of the support.

5) *Differential migration* begins: each component migrates up the support by a certain height, characteristic of the substance being analyzed.

6) *Comparative analysis*: comparison of the migration of the various constituents of the mixture with reference chemicals, also known as controls. This makes it possible to separate the constituents of the mixture.

Figure 4.5(b) shows the chromatography cell. It shows the *baseline* or *deposition line* and the *solvent front*, two fundamental concepts used to determine RCP.

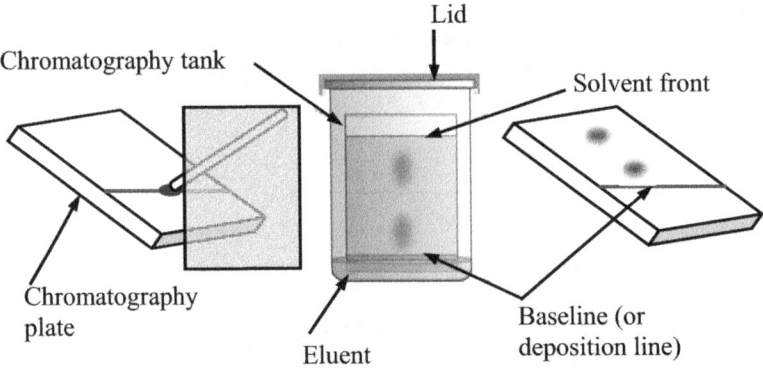

Figure 4.5(b). *Chromatography tank. Here the solvent front and the deposition line can be seen*

The experimental protocol for TLC can be summarized in five steps [MAG 21].

Step 1: choice of support (fixed phase) and eluent (mobile phase):

– Choose the fixed phase (aluminum plate covered with silica gel).

– Choose the mobile phase. Solvent or solvent mixture (dichloromethane, petroleum ether, etc.).

Step 2: preparing the chromatography tank and support:

– Pour approximately 0.5–1 cm of eluent into the chromatography tank, which is closed with a lid so that the eluent saturates the tank with vapor.

– Draw a thin line called the *deposition line* (or *baseline*) on the chromatography plate so that it is above the level of the eluent.

Step 3: deposit preparation:

– Make the various deposits on the base line: mixing.

– Choose reference or control chemicals (products likely to be used in the composition of the mixture).

– Respect regular spaces between each deposit: place fine crosses at the location of completed deposits to be identified by a letter or a name.

– Finally, these deposits must be dried to fix them firmly to the substrate.

Figure 4.5(c) shows how to prepare the baseline.

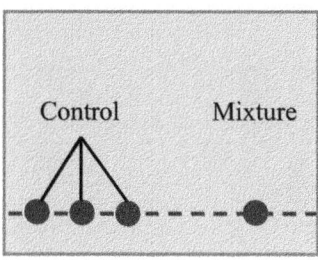

Figure 4.5(c). *Baseline preparation principle*

Step 4: performing chromatography:

To do this:

– Place the plate vertically in the tank and replace the lid.

– Allow the eluent to migrate by capillary action.

– Remove the plate when the solvent reaches ~0.5 cm from the top of the plate, tracing a new line called the *solvent front*; the stain made up of the mixture will migrate upwards, dividing into as many stains as there are constituents.

Step 5: revealing and analyzing the chromatogram by comparison:

When performing TLC, staining spots are not necessarily visible. Sometimes, it is necessary to dip the plate in a developer to make them visible. Stains that have reached the same height are made up of the same product.

Therefore, during chromatography, the chemical species analyzed migrate at the same speed and appear at the same height on the chromatogram, making it possible to derive a great deal of information from a chromatogram:

– The substance analyzed presents a single stain: only one chemical species is present (it is a pure body).

– The substance analyzed has several stains: we are dealing with a mixture of several different chemical species.

– The substance analyzed shows a stain at the same level as a reference chemical species (control): the presence of this species in the substance analyzed is then confirmed (Figure 4.5(d)).

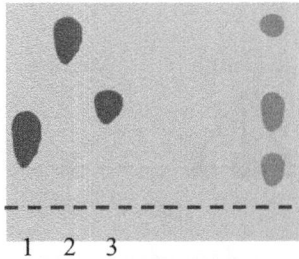

Figure 4.5(d). *Example of chromatogram. Only the components of controls 2 and 3 are present in the mixture. The component of control no. 1 is absent, since the corresponding stain is not at the level of any stain from the mixture.*

Figure 4.5(d) shows an example of the end of chromatography with controls numbered 1–3. The chromatogram shows that only the constituents of controls 2 and 3 are present in the mixture. The constituent of control no. 1 is not present in the mixture, as the corresponding stain is not at the level of any stain from the mixture.

– Frontal ratio determination in chromatography:

In chromatography, the *frontal ratio* (R_f) is a physical quantity used to characterize a given chemical species for a given solvent and fixed phase. By calculating the frontal ratio, we can determine the nature of an unknown chemical species.

By definition, R_f is equal to the ratio of the distance h traveled by a deposited chemical species from the deposition line to its final position on the chromatogram to the distance H separating the deposition line from the eluent front (Figure 4.5(e)), i.e.:

$$R_f = h / H \qquad [4.2c]$$

Knowing that h and H are expressed in the same unit (usually both in centimeters or millimeters), then the frontal ratio R_f is a unitless quantity.

As shown in Figure 4.5(e), h is less than H. Consequently, the frontal ratio theoretically verifies the double inequality: $0 \leq R_f \leq 1$.

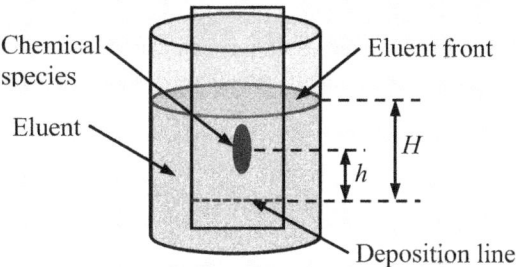

Figure 4.5(e). *Principle for determining the frontal ratio on a chromatogram. The frontal ratio of the chemical species analyzed $R_f = h/H$*

4.3.5. Determination of radionuclidic purity

As defined above, radionuclide purity is the ratio of the radioactivity of the isotope in question to the total radioactivity of the radiopharmaceutical. Its experimental measurement enables us to determine the nature and energy of the radiation emitted by the RPP, and to identify the radionuclide by measuring its physical half-life (radioactive decay curve). As an illustrative example, let us consider the case of ^{18}FDG, a positron emitter, whose annihilation produces gamma photons of 511 keV.

– Gamma-ray spectroscopy is used to determine the nature and energy of the radiation emitted:

In practice, a compliance test must be carried out before using the device. This is carried out using a cesium source. Once this test has been carried out, the nucleic acid purity of the solution can be checked. A peak at 511 keV ± 5% (385–537 keV) corresponding to the energy of the gamma photons must be observed to confirm that the radionuclide present is ^{18}F (measurement of the half-life will enable conclusions to be drawn). Another gamma spectrum is taken after decay (at least 23 hours) to ensure the decay of the ^{18}F. During this test, fluorine-18 and "long-lived" radionuclide impurities are quantified [BOU 17].

– Radionuclide identification: determination of half-life:

The radionuclide can be identified by measuring its physical half-life (radioactive decay curve). For ^{18}F, the half-life is 109.8 minutes. This measurement is made using an activimeter, by taking at least three measurements of the activity of a sample under the same geometric conditions and over an appropriate period of time. In practice, a conformity test must be carried out before using the device. This is carried out using a cesium source. Once this test has been carried out, and if it is satisfactory, the period can be measured. This is calculated using an algorithm, with measurements taken every minute for 20 minutes. It should be between 105 and 115 min [BOU 17].

4.4. Nuclear medicine imaging techniques: PET and SPECT

Nuclear medicine imaging uses two different diagnostic techniques: *positron emission tomography* (PET) and *single-photon emission computed tomography* (SPECT). While SPECT uses gamma-photon emitting radiopharmaceuticals, PET uses positron-emitting radiopharmaceuticals (β^+). After annihilation with negative electrons, gamma photons are emitted into the biological medium under consideration. In this section, we will confine ourselves to presenting the principles of PET and SPECT. A detailed study of these two techniques will be covered in other volumes.

4.4.1. Radioisotopes used in nuclear medicine imaging

Nuclear medicine imaging is a medical imaging method based on the injection of a radiotracer into the body. Most often, the radiotracer is injected intravenously at very low doses (10^{-12} mol). It is thus designed to bind (metabolize) specifically to the organ to be imaged, emitting photons that pass through the patient's body.

Nuclear imaging is used to diagnose three major types of disease: cancer, cardiovascular disease and neurological disorders such as Alzheimer's disease [FRA 14]. Two different techniques are used for these diagnoses: *single-photon emission computed tomography* (SPECT) and *positron emission tomography* (PET). Imaging accounts for 90% of the use of all medical radioisotopes [COM 16].

Note that cardiovascular diseases constitute a group of disorders affecting the heart and blood vessels, which includes: coronary heart disease (affecting the blood vessels that supply the heart muscle), and cerebrovascular diseases (affecting the blood vessels that supply the brain).

Various radionuclides used in SPECT are presented in Table 4.3 [FRA 14].

Single-photon emission computed tomography (SPECT)				
Radioisotope	Radiation	Half-life	Production	E_{max} (keV)
^{131}I	γ, β^-	8.0 d	^{130}Te/^{131}I reactor	333, 606
^{131}In	γ	67.2 h	^{111}Cd/^{111}In cyclotron	171, 235
^{123}I	γ	13.2 h	^{127}I/^{123}Xe cyclotron	159
^{99m}Tc	γ	6.02 h	^{99}Mo/^{99m}Tc generator	130
^{201}Tl	γ, β^-	72.96 h	^{201}Pb/^{201}Tl generator	135

Table 4.3. *Radioisotopes used in SPECT*

Unlike SPECT, where the radiopharmaceuticals used are gamma photon emitters, PET uses positron-emitting radiopharmaceuticals (β^+ particles). However, it should be noted that the range of radiopharmaceuticals available in SPECT is much wider. This nuclear medicine imaging method, which is much simpler to use than PET, remains irreplaceable for many applications. The main radioelements used in PET are listed in Table 4.4 [ZIM 06, JOY 13, GAS 18].

Positron emission tomography (PET)				
Radioisotope (^{A}X)	Period (T)	β_{max}^+ (MeV)	β^+ (%)	Production
Rubidium-82m (^{82m}Ru)	6.3 h	3.15	100	Strontium-82 generator
Carbon-11 (^{11}C)	20.3 min	0.96	100	Cyclotron
Gallium-68 (^{68}Ga)	1.12 h	1.90	89	Germanium-68 generator
Fluorine-18 (^{18}F)	108 min	0.63	97	Cyclotron
Copper-63 (^{63}Cu)	12.7 h	0.66	18	Cyclotron
Yttrium-86 (^{86}Y)	13.7 h	1.25	33	Cyclotron
Zirconium-89 (^{89}Zr)	78.3 h	0.90	23	Cyclotron
Iodine-123 (^{123}I)	3.18 d	2.13	23	Cyclotron

Table 4.4. *Main radioelements used in PET*

Among the radioisotopes listed in Tables 4.3 and 4.4, fluorine-18, technetium-99m, iodine-123 and iodine-131, as well as other radioisotopes such as thallium-201, are the most commonly used. Fluorine-18 has been identified as the ideal radioisotope for PET.

4.4.2. Principle of positron emission tomography (PET)

PET is an emission tomography using a radiopharmaceutical [TAL 03, VAL 09, NIC 10, GAS 18, MAN 20, SAL 20]. Once created, the positron encounters an electron in matter and an annihilation process occurs. This produces the simultaneous emission of two γ photons propagating in two opposite directions (Figure 4.6). Furthermore, these photons have the same energy of 511 keV, whatever the β^+ emitter.

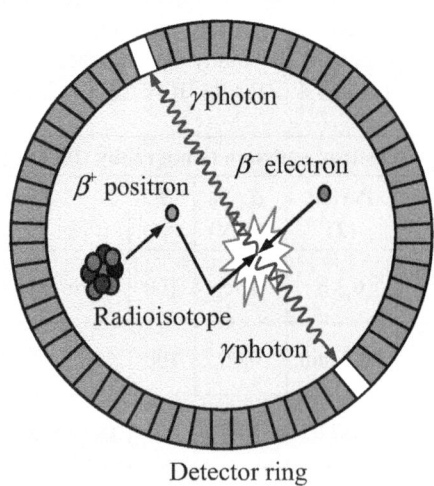

Detector ring

Figure 4.6. *Process of annihilation of a positron emitted by a radioisotope*

The principle of PET is to place suitable sensors on either side of the emission site, coupled with suitable computers. This makes it possible to locate the point of origin of the shock between the positron and the electron encountered. Imaging analysis techniques, combined with data recording in successive slices, can be used to create 2D or even 3D images. Figure 4.7 summarizes the principle of detecting γ photons resulting from the annihilation of a positron and an electron, using detectors placed in a ring around the patient.

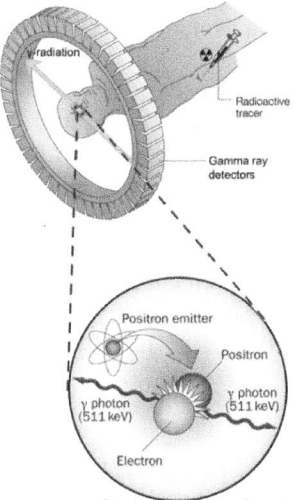

Figure 4.7. *Principle of positron emission tomography (PET)*
(see: https://multimodalneuroimaging.wordpress.com/
2015/04/28/c-positron-emission-tomography/)

Note that, depending on the positron's ejection energy, it can travel a distance of several millimeters from its point of ejection before encountering an electron. The final image, corresponding to the sum of the impact points, will show the statistical distribution of these annihilation points, rather than the distribution of the origin of the positron emission. These few millimeters of difference with the real origin of the β^+ signal also express the inescapable limit of resolution and therefore image quality of the method.

This resolution, of the order of a few millimeters, is nevertheless considered excellent, thanks to the specificity of the vectors used. Some micro-metastases, i.e. metastases and tumors of the order of 3–5 millimeters in diameter, are still visible due to the high contrast with the background, but their actual size cannot be assessed. This property of imaging very small areas is therefore not linked to the radioisotope, but to the specificity of the vector onto which this element is grafted [ZIM 06].

4.4.3. *PET scan*

A PET scan is a nuclear medicine imaging technique increasingly used to study organ function, metabolism and anatomy [GIO 19, CEN 21]. It is used in oncology, cardiology and neurology. This medical imaging can detect malignant tumors and their metastases, and monitor their evolution.

In oncology, the concentration of tracer in certain areas reveals the presence, activity and extent of cancerous tumors. It can also be carried out repeatedly to monitor the effectiveness of treatment, or to search for any metastases invisible on an image obtained using another imaging technique. This examination is also used in cardiology to analyze blood flow in coronary arteries or heart chambers, and to visualize the extent of lesions following a myocardial infarction. In neurology, it is used to assess brain function in certain neurodegenerative pathologies such as Alzheimer's or Parkinson's disease. Finally, it can detect certain abnormalities inaccessible to other imaging techniques [GIO 19].

The principle of this technique is identical to that of PET, except that it is coupled with atomic imaging using a scanner. The principle involves injecting a weakly radioactive product into a patient, which then binds to tumor cells and/or metastases. Due to its low-level radioactivity, it can then be monitored using a PET scan, for which the most widely used product is ^{18}FDG [DEB 08].

It should be noted that a PET scan can be combined with a CT scan to give an accurate picture of both metabolism and anatomy. While PET scans are highly sensitive to increased metabolic activity, their images are less precise than those of CT scans. For this reason, many hospitals have hybrid machines capable of combining conventional CT scan and PET scan. By comparing the results of both examinations, doctors can locate tumors and metastases with great precision [FON 20].

NOTE.–

PET/CT stands for positron emission tomography/computed tomography. The French equivalent is TEP/TDM Tomographie par Emission de Positons/ Tomodensitométrie (TomoDensitoMétrie).

4.4.4. *PET scan procedure*

Apart from the requirement to fast (water and unsweetened beverages may be consumed, but no meals), the PET scan requires no special preparation. The procedure is as follows:

– The patient should not have eaten for at least six hours.

– Blood sugar levels are checked on arrival at the nuclear medicine department.

– When the blood glucose levels are correct (between 0.7 g/L and 1 g/L), the tracer is injected.

– The patient lies on the PET scan examination table (Figure 4.7) for about one hour, to allow the drug to distribute itself throughout the body and be taken up by the sugar-consuming cells.

– The patient goes under the camera for about 10 minutes.

– PET scan sensors record the radioactive activity of the injected product. Once the examination is complete, powerful computers reconstruct the final images from the recorded data.

The isotope injected is eliminated by the body within a few hours. Furthermore, the examination has no known long-term side effects, since the quantities of drug injected are very small. There are no allergic reactions, nausea or vomiting [GIO 19].

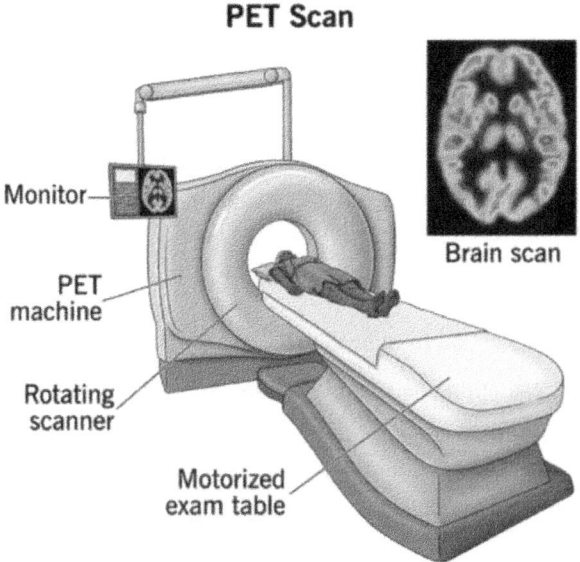

PET Scan

Monitor

PET machine

Rotating scanner

Motorized exam table

Brain scan

Figure 4.8. *PET scan equipment. See: https://my. clevelandclinic.org/health/diagnostics/10123-pet-scan*

In addition, blood glucose levels should be measured prior to administration. Indeed, hyperglycemia, especially when it exceeds 8 mmol/L, can reduce the sensitivity of a PET scan with FDG scan (^{18}F) 260 MBq/mL [AGE 09]. This is not a contraindication, but it is important to point this out [BOU 17], or simply to avoid administering FDG scan (^{18}F) in subjects with unbalanced diabetes [AGE 09].

The PET scan (Figure 4.8) looks like a scanner, but functions differently, as it uses the scintigraphy method. It consists of:

– a detector for gamma rays;

– an X-ray scanner;

– a table on which the patient lies down and which moves towards the ring.

By combining the two images, we can optimize image quality and better visualize the organs where an anomaly has been detected.

NOTE ON BLOOD SUGAR.–

– Normal fasting blood glucose ranges from 0.7 g/L and 1 g/L. Below a certain threshold, we speak of hypoglycemia, whereas above, we speak of hyperglycemia. Diabetes, on the other hand, is characterized by fasting blood glucose levels of over 1.26 g/L.

– Glycemia is the level of glucose (sugar) in the blood. Sugar is one of the nutrients essential to the proper functioning of the body's cells, and is used for energy production. In fact, some of the glucose in the blood is converted into glycogen, which supplies the body with energy [THI 20].

Good to know [FON 20]:

– PET scans analyze you from head to toe in a single examination.

– Excluding preparation, the exam lasts approximately 50 minutes.

– PET scans have their limitations, however, and can give false negatives (i.e. it does not always detect existing tumors) for the following reasons:

  - Some tumors do not bind the isotope and therefore go undetected.

  - The isotope does not reach certain parts of the body, so the tumor cannot be visualized.

  - Although the resolution of PET scans is constantly improving, very small tumors remain undetectable. These small tumors bind the isotope correctly, but their mass is less than the minimum size of the elements visible on the image. This is why tumors smaller than approximately 5–10 mm are difficult to detect on a PET scan.

– PET scans can also produce false positives, detecting a "tumor" where none exists.

4.4.5. *PET/CT examination*

As noted above, a PET scan can be combined with a CT scan. This fusion of the two medical imaging techniques provides images of both the anatomy (using the CT part of the scan) and the metabolic activity of the body's cells (using the PET part of the scan).

The PET/CT examination begins with an IV injection of ^{18}FDG. This radiopharmaceutical is then absorbed by the body's various cells, which use glucose as their main source of energy. As a result, tumor (cancer) cells with increased metabolism will accumulate more ^{18}FDG than healthy cells.

One hour after injection (sufficient time to allow cancer cells to accumulate sufficient ^{18}FDG), PET/CT images are taken. In practice, we start with the CT examination, immediately followed by PET imaging. Subsequently, the two sets of images (CT and PET) obtained are merged for precise localization of areas of ^{18}FDG hyperfixation [CSM 21].

PET-CT is often used for research into the functioning of neurodegenerative diseases such as Alzheimer's disease and Parkinson's disease. Today, new studies have demonstrated the value of using PET to diagnose Alzheimer's disease before it develops. However, other imaging tests such as MRI (PET/MRI) or CT (PET/CT) are required to complete the picture.

4.4.6. *Principle of single-photon emission computed tomography*

Scintigraphy is also known as *diagnostic imaging* or single-photon emission computed tomography (SPECT). For this reason, SPECT (single-photon emission computed tomography) is based on the principle of *scintigraphy* [BEK 06, DEB 08, LAP 12, ANS 16, POI 19, MAN 20]. Injected at very low doses (10^{-12} mol), the radiotracer does not cause side effects or allergies [HAN 16]. It is designed to bind (metabolize) specifically to the organ to be visualized, emitting photons that pass through the patient's body. These photons are detected by a special camera called a "gamma camera". This camera revolves around the patient's body, and the images thus collected are analyzed and collated to form a 3D image of the organ under study. The image obtained is very detailed. It should be noted that in some cases, the tracer is used by the cells as a source of energy, enabling areas of higher cellular activity to be identified (e.g. the brain or heart, or a tumor). SPECT has applications in the exploration of the heart, brain, lungs and skeleton.

A description of the tomoscintigraphy technique and its advantages and disadvantages are summarized below (source: https://www.larousse.fr/encyclopedie/medical/tomoscintigraphie/16612).

– *Description*:

The most commonly used device is a tomograph, consisting of a gamma camera (a special camera sensitive to gamma radiation) attached to a support capable of revolving around the patient's body, thus recording images from multiple angles. The computer associated with the gamma camera determines the radioactivity contained in each elementary volume, or voxel, of the organ studied. The patient, to whom the radioactive tracer has been injected intravenously, absorbed or inhaled, is in the lying position. The gamma camera detector is placed in front of the organ to be studied, perpendicular to the main axis of the body, and then begins to rotate slowly around this axis in either a continuous or fractionated ("step-by-step") movement. Depending on the case, the rotation may be complete or consist of only a half-turn. It lasts between 10 and 20 minutes, depending on the amount of radioactivity concentrated in the region under study. At the end of the recording phase (known as the "acquisition" phase), the information entered in the computer memory is processed to provide, after calculation, cross-sectional images.

– *Advantages and disadvantages of tomoscintigraphy*:

Compared with conventional scintigraphy images, which provide a two-dimensional representation of the structure under study, tomoscintigraphy cross-sectional images enable tracer distribution anomalies to be identified with greater precision. Furthermore, they can be reconstructed from different angles, or even combined with X-ray CT images, to help pinpoint anomalies. The main drawback of this technique lies in the difficulty of studying a phenomenon when it evolves too rapidly. In the case of the heart, however, this difficulty can be overcome by synchronizing images with cardiac contractions, using an electrocardiogram.

4.4.7. *Main scintigraphies and their uses*

Scintigraphy is generally used to explore various organs (thyroid, myocardium, bone, kidney, lung, brain, etc.). Table 4.5 shows the main scintigraphies and their uses in nuclear medicine.

*Coronary insufficiency (coronary artery disease). This is a disease of the arteries that vascularize the heart (coronary arteries), resulting in myocardial ischemia, i.e. insufficient blood supply (ischemia) to the heart muscle (myocardium).

Main scintigraphies	Uses
Cardiac scintigraphy	Diagnosis of coronary insufficiency*, post-myocardial infarction assessment.
Bone scintigraphy	Detection of bone metastases, bone diseases (e.g. Paget's disease), stress fractures, etc.
Pulmonary scintigraphy	Diagnosis of pulmonary embolism, follow-up of chronic bronchial disease.
Renal scintigraphy	Diagnosis of kidney malformation or dysfunction, assessment in case of tumor or kidney infection (e.g. pyelonephritis).
Thyroid scintigraphy	Diagnosis of hyperthyroidism, Graves' disease, nodules, thyroiditis, etc.

Table 4.5. *Main scintigraphies and their uses. Source : https://www.vidal.fr/ sante/examens-tests-analyses-medicales/scintigraphie- tomographie-emission-monophotonique.html 2017*

4.5. Appendices on dementia diseases

The concept of dementia dates back to antiquity [ROU 17]. Originally a synonym for *madness*, the term encompasses loss of mind or intelligence and extravagant behavior beyond social norms (delusions, manias, etc.). The main facts, signs and symptoms, common forms, statistics, treatment and care as well as risk factors and prevention relating to dementia are presented below, according to the WHO (World Health Organization) [WHO 23].

In section 3.4.4 of Chapter 3 of Volume 3, we will examine the properties of the radiopharmaceutical ^{123}I-ioflupane. This ^{123}I-ioflupane radiopharmaceutical (Datscan®) is used in practice for differential diagnosis between essential tremor (a non-neurodegenerative disease, see section 3.4.5) and neurodegenerative diseases such as Parkinson's disease, Lewy body dementia and Alzheimer's disease. In this appendix, we look in detail at the main features of Parkinson's disease, Alzheimer's disease and Lewy body dementia.

4.5.1. *Appendix 1. Alzheimer's disease*

4.5.1.1. *General information*

In 1975, Turkish psychiatrist, psychoanalyst and writer Toksöz Bayram Katzman (1925–2008) and American neurologist Robert Karasu (b. 1935) concluded that cases of senile dementia should be included in the diagnosis of Alzheimer's disease.

As a result, the first world congress on "Alzheimer's dementia and other senile dementias" was held in London in 1977 [ROU 17].

Alzheimer's disease is a neurodegenerative disorder responsible for cognitive and behavioral disorders. It is a brain disease leading to the progressive and irreversible loss of mental functions, particularly memory. The disease was first described by Alois Alzheimer in 1906 (see the note). It should be noted that a neurodegenerative disease is a pathology leading to the death of neurons and destruction of the nervous system [DOM 18].

According to the 2015 report by ADI (Alzheimer's Disease International), Alzheimer's disease is the first cause of dementia: 60–70% of cases. The number of new cases of dementia in 2015: 9.9 million, or one new case every three seconds. The breakdown of new cases in 2015 was as follows [ALZ 15]:

– 49% in Asia;

– 25% in Europe (down on 2012 figures);

– 18% in America;

– 8% in Africa.

The number of cases of dementia will almost double every 20 years.

4.5.1.2. *Causes and effects, risk factors, diagnosis*

Alzheimer's disease results from a slow degeneration of neurons, beginning in the hippocampus (a brain structure essential for short-term memory). Several brain regions are then affected, and the disease gradually spreads to the whole brain.

The main risk factor for Alzheimer's disease is age. Genetic factors are also involved. Several genes are thought to be linked to increased susceptibility to the disease (amyloid peptide metabolism genes, genes involved in inflammation, genes involved in neuronal communication, etc.). Conversely, certain genes appear to protect against the disease. The environment also seems to play a role. The disease may be more prevalent in people who are sedentary, have undergone repeated anesthesia or are exposed to unmanaged cardiovascular risk factors (diabetes, hypertension, etc.). Faced with the inescapable, there is a "cognitive reserve", on which the environment has a positive impact. In this way, the function of lost neurons can be compensated for by stimulating the brain, and the onset of the first symptoms and/or their severity can be delayed by continuing education, a stimulating professional activity, or an active social life, etc.

– Diagnosis:

While there is no cure for Alzheimer's disease, its progression can be slowed down. That is why early diagnosis is so important, not only for the rapid implementation of current drug treatments, but also for the application of current preventive advice aimed at stimulating memory capacity. Early diagnosis also enables the patient to make arrangements for their future life while they still *possess all the faculties of discernment and decision-making.* Source: https://www.pasteur.fr/ fr/centre-medical/fiches-maladies/alzheimer-maladie, 2021.

4.5.1.3. *Alzheimer's disease worldwide*

The global state of Alzheimer's disease in 2019 is summarized in the work of Olivier de Ladouchette, President of the Alzheimer's Research Foundation (Les chiffres de la maladie d'Alzheimer - Aidons les nôtres (aidonslesnotres.fr):

> More than 50 million, or about one in every 20 elderly people! That's the number of individuals suffering from dementia worldwide, according to a study commissioned by Alzheimer's Disease International (ADI) in 2019.

> Given that Alzheimer's disease is the most common cause of dementia, accounting for 60–70% of all cases, it can be considered to affect around 37 million people. The disease does not affect all parts of the world in the same way: Asia leads the way with almost half of all cases. Europe follows with a quarter of cases, while America accounts for just 18% and Africa for 8%. Logically, the more a continent's population ages, the more it is affected by this age-related disease.

> – Alzheimer's disease in France

> Around 900,000 French people suffer from Alzheimer's disease, according to the Inserm information file dedicated to this pathology (2014). More than 225,000 new cases are recorded every year. Most of these are elderly people (20% of those over 80 suffer from the disease), but the number of patients under 60 is 33,000. Women are more affected than men (60% of cases).

The state of Alzheimer's disease in the UK, USA, Quebec and Senegal is as follows.

– Alzheimer's disease in the United Kingdom:

The Office for National Statistics (ONS) has published its first-ever figures relating to dementia and deaths from Alzheimer's disease, which also include comorbidities, for England and Wales. The researchers found that in 2019, 530,841 deaths were recorded. Of these, there were 66,424 deaths due to dementia and Alzheimer's disease (12.5%), compared with 69,478 deaths in 2018. The ONS also noted a "significant decrease" in the age-standardized mortality rate in 2019 compared with 2018 (a fall from 123.8 per 100,000 people in 2018 to 115.1 per 100,000 people in 2019).

There are four main causes of death in the UK: dementia (including Alzheimer's disease), heart disease, lung cancer and stroke. Statistics showed that dementia was the biggest killer in the UK for the first time in 2015. Although deaths related to Covid-19 were the leading cause of death in England and Wales in 2020 and 2021, dementia was again listed as the leading cause of death in 2022. In these figures, all diseases that cause dementia are considered together, whereas the various types of cancer are separated. However, Alzheimer's disease alone would make up an estimated 13% of deaths. See: https://www.alzheimers.org.uk/blog/research-UK-biggest-killer-high-dementia-deaths.

– Alzheimer's disease in the USA:

According to Agence France Presse [AGE 13a], one in three elderly people in the United States dies with Alzheimer's disease or another form of dementia, as reported by the American Alzheimer's Association on Tuesday. "Today there is no Alzheimer's survivor, either you die of it or you die with it," notes Harry Johns, president of the Alzheimer's Association. *"We now know that one in three older people dies with Alzheimer's or other dementias, and there is an urgent need to act as, with an aging population, more and more people are at risk of falling victim to this incurable brain degeneration with no treatment to slow or halt its progression."*

Alzheimer's disease is the sixth leading cause of death in the United States, with 83,494 deaths in 2010. This represents an increase of 39% over 10 years, according to the latest federal statistics. According to figures from the American Alzheimer's Association, 450,000 people will die with the disease in 2013, a significant proportion of them as a direct result of this chronic pathology. Even when Alzheimer's disease is not the direct cause of death, it contributes significantly to it, as the authors of the study point out. Recent research on this subject, based on a national sample, has shown that dementia is the second most important mortality factor after heart failure. Therefore, among septuagenarians suffering from Alzheimer's, the probability of dying within 10 years is 61%, compared with only 30% for those not suffering from this disease, says the study.

Alzheimer's affects 36 million people worldwide, including 5.5 million Americans. This number could reach 13.8 million by 2050 in the United States. The cost of the healthcare system is high. It was expected to reach $203 billion in 2013, including $142 billion for Medicare, the federal health insurance for retirees, and Medicaid, the coverage for the poorest, according to an estimate by the American Alzheimer Association. Without medical advances against the disease, this cost is expected to rise by 500% by 2050 to $1,200 billion, as the association predicts. According to UsAgainstAlzheimer's, a private foundation at the forefront of the fight against the disease, the United States should devote 2 billion dollars a year to research to combat Alzheimer's, 4.4 times more than the current budget.

– Alzheimer's disease in Quebec:

The figures for Quebec are as follows [LAS 17-21]:

– There are 1,692,486 people over 65 in Quebec.

– There are 152,121 people with Alzheimer's disease or a related disorder in Quebec.

– There are 15,088 people with Alzheimer's disease or a related disorder in the Capitale-Nationale region (Charlevoix, Portneuf, Quebec).

– Alzheimer's disease is the most common form of neurocognitive disorder, accounting for more than 60% of cases.

– Within 15 years, these diseases will affect 937,000 people in Canada and 260,000 in Quebec, an increase of 66%.

– The costs associated with Alzheimer's and other neurodegenerative diseases represent $2 million and are expected to reach $123 billion by 2040.

– Nearly 50% of people with cognitive disorders, including Alzheimer's disease, are diagnosed at an advanced stage of the disease. As it is progressive, this progression varies from person to person, and can last 8–10 years or more. Cognitive diseases can lie dormant in the brain for up to 25 years before symptoms appear.

– For each person affected, there are one to three caregivers who invest time and care, with financial and socio-economic impacts on the immediate family, the job market and our society.

– Women account for 70% of all caregivers.

– 65% of Canadians affected are women.

– After the age of 65, the risk of developing Alzheimer's disease doubles every five years. Age is the biggest risk factor, and the disease strikes people in their 50s, 40s and even 30s.

– 565,000 Canadians, including 152,121 Quebecers, live with Alzheimer's disease or a related dementia, and 15,088 people in the province of Quebec.

– *Alzheimer's disease in Senegal*:

Until 2019, the number of individuals suffering from dementia was estimated at 8% in Africa. In the specific case of Senegal, it is estimated that [SEN 21] by the year 2028, 9% of the Senegalese population, i.e. approximately 1.5 million Senegalese, will be affected by Alzheimer's disease or the disease of forgetting, according to a study carried out by the Institution de prévoyance retraite du Sénégal (I.P.RE.S.). The age group 85 and over will have a 22% incidence of Alzheimer's disease, which to-date remains incurable.

By 2022, it is estimated that 900,000 people in France will be suffering from Alzheimer's-type dementia, and 35 million worldwide. While the onset of dementia before the age of 65 is rare (0.5%), the incidence rises to between 2% and 4% after this age. It then increases proportionately with age, reaching over 15% at 80. The disease is increasingly affecting women (one in four women and one in five men over the age of 85 years) [INS 22].

NOTE.–

It is worth recounting the story of Madame Auguste Deter (1850–1906), the first person to be diagnosed with Alzheimer's disease.

In the late 1890s, Madame Auguste Deter, wife of Carl (Karl) August Wilhelm Deter, began to show symptoms of dementia, such as memory loss, delirium and even temporary vegetative states. She would have trouble sleeping, dragging sheets around the house and screaming for hours in the middle of the night.

It should be noted that a vegetative state should not be understood in the botanical sense, where it means one of the stages in the growth of a plant. In the case of dementia, a vegetative state is the state of a person who is apathetic (lacking energy or emotional reactivity), seemingly no longer interacting with their environment, their higher brain functions not being mobilized at any time.

As a railway worker, Carl was unable to provide adequate care for his wife. He had her admitted to a psychiatric institution, the Institution for the Mentally Ill and Epileptics in Frankfurt, Germany on November 25, 1901. There, she was examined by Dr. Alois Alzheimer.

– *The treatment*:

Dr. Alzheimer asked her many questions, then asked her again if she remembered. He told her to write her name. She tried, but forgot the rest and repeated, "I've lost myself." (German: "Ich habe mich verloren.") Later, he put her in solitary confinement for a while. When he released her, she would run screaming, "I will not be cut. I'm not cutting myself."

Alzheimer concluded that she had no idea of time or place. She could barely remember the details of her life and frequently gave answers that had nothing to do with the question and were incoherent. Her moods changed rapidly between anxiety, distrust, withdrawal and "whining". They could not let her wander the wards because she would approach other patients who would then assault her. This was not the first time Dr. Alzheimer had seen complete degeneration of the psyche in patients, but previously, patients had been in their 70s. Mrs. Deter piqued his curiosity because she was so much younger. In the weeks that followed, he continued to question her and record her answers. She often replied, "Oh, my God!" and "I've lost myself, so to speak." She seemed aware of her helplessness. Alzheimer's called it the "disease of forgetting".

– Death and legacy:

In 1902, Alzheimer left the "Irrenschloss" (Castle of Fools), as it was familiarly known, to take up a position in Munich, but made frequent calls to Frankfurt to enquire about Deter's condition. On April 9, 1906, Alzheimer received a call from Frankfurt stating that Auguste Deter had died. He requested that her medical file and brain be sent to him. Her file indicated that in the last years of her life, her condition had deteriorated considerably. Her death was the result of a septic condition caused by an infected eschar (skin lesion). Examination of her brain revealed senile plaques and neurofibrillary tangles. This would be the hallmark of Alzheimer's disease as scientists know it today. Auguste would have been diagnosed with early-onset Alzheimer's disease had some of today's doctors seen her.

– Rediscovering the medical record:

In 1996, Dr. Konrad Maurer and his colleagues, Drs. Volk and Gerbaldo, rediscovered Auguste Deter's medical records. In these documents, Dr. Alzheimer had recorded his examination of his patient, including her answers to his questions (*we have put Mrs. Auguste Deter's answers in bold to distinguish them from Alzheimer's questions*):

"What's your name?"
*"**Auguste**."*
"Last name?"
*"**Auguste**."*
"What is your husband's name?" - she hesitates, finally answers:
*"**I believe… Auguste**."*

"Your husband?"
"Oh, my husband."
"How old are you?"
"Fifty-one."
"Where do you live?"
"Oh, you've been to our house."
"Are you married?"
"Oh, I'm so confused."
"Where are you right now?"
"Here and everywhere, here and now, it's best not to think about me too much."
"Where are you at this moment?"
"We'll live there."
"Where is your bed?"
"Where should it be?"

Around midday, Frau Auguste Deter (Mrs. Auguste Deter) ate pork and cauliflower:

"What are you eating?"
"Spinach." (She was chewing meat.)
"What are you eating now?"
"I eat potatoes first and then horseradish."
"Write a '5'."
She writes: **"A woman"**
"Write an '8'."
She writes: **"Auguste"** (As she writes, she repeatedly says: "I've lost myself, so to speak").

Alois Alzheimer (1864–1915) was a German psychiatrist and neurologist. He is best known for his 1906 description of the dementia known as Alzheimer's disease, named in his honor. He followed the dementia-stricken Madame Auguste Deter in 1901. After her death in 1906, Alzheimer carried out a brain autopsy in Munich, which revealed the anomalies that would become characteristic of Alzheimer's disease.

4.5.2. Appendix 2. Lewy body dementia

4.5.2.1. Definitions, mixed dementia

Lewy body disease (LBD), also known as *Lewy body dementia* (LBD), is a complex neurodegenerative disease in which any part of the brain can be affected. Although less well-known than Alzheimer's disease (AD) or Parkinson's disease (PD), LBD is nonetheless very common, accounting for approximately 20% of all dementias. This type of dementia is named in honor of the German neuroanatomist and psychiatrist Friedrich Heinrich Lewy (1885–1950). In 1912, he discovered the cellular inclusions (see the note) of protein substance observed in certain nervous system pathologies, known as *Lewy bodies*.

Lewy bodies are abnormal aggregates of proteins that form inside nerve cells during Parkinson's disease, Lewy body disease and certain other neurodegenerative diseases. This justifies the frequent confusion between LBD and PD. However, it is possible to differentiate between them on the basis of how the disease starts. Parkinsonian dementia is characterized by movement disorders, whereas Lewy body dementia most often begins with cognitive disorders. However, when the cognitive symptoms of LBD appear long after the parkinsonian symptoms, the dementia is considered to be parkinsonian in nature.

Clinical signs depend on the location of the lesions. LBD generally begins after the age of 50. It seems to affect men slightly more than women. The course of the disease is highly variable. Life expectancy from diagnosis can vary from 2 to 20 years. Clinical signs depend on the location of the lesions. There is no cure, but symptomatic treatments are available [UNI 18].

VOCABULARY CORNER.–

Cellular inclusions (or cytoplasmic inclusions) are non-living intracellular substances that do not carry out any type of biochemical reaction. Note that cytoplasm refers to the contents of a living cell.

Neuropathological research has demonstrated the existence of mixed dementias, characterized by the concomitance of several causes of dementia. In an autoptic study, it was shown that half of all people with dementia had several associated causes: Alzheimer's dementia and vascular dementia (38%), Alzheimer's dementia and Lewy body dementia (12%), or a combination of all three (2%) [SCH 07]. The association of vascular dementia with Alzheimer's disease in the context of mixed dementia is frequent: approximately 50% of patients with Alzheimer's dementia have anatomopathological evidence of silent infarction [FER 04].

4.5.2.2. *Symptoms, diagnosis and prevention of Lewy body disease*

The symptoms, diagnosis and medications recommended for the prevention of LBD are summarized below, according to [UNI 18]:

– *Symptoms*:

  - *Cognitive symptoms*:

Loss of cognitive ability is often one of the first symptoms of LBD.

Generally speaking, sufferers experience difficulties with visual and spatial perception. Attention disorders, very common at the onset of the disease, may be confused with memory problems. They may also have difficulty performing several tasks simultaneously, or reasoning logically.

Unlike Alzheimer's disease, memory problems may not appear at the onset of the disease. Instead, they often manifest themselves later, as the disease progresses. The sufferer may also experience changes in mood and behavior that may be reminiscent of depression.

- Cognitive fluctuations:

Cognitive fluctuations are characteristic of the disease, as are unpredictable changes in concentration, attention, alertness and wakefulness. These changes can occur from one day to the next, or even from one hour to the next. A person with LBD may stare into space for long periods, or appear drowsy and lethargic, spending a lot of time asleep. Ideas may be confused, without apparent logic or seemingly random. Speech may be nothing more than "word salad". However, at other times, the same person will be alert, able to carry on a lucid conversation, laugh at a joke or even follow a movie. Although these fluctuations are common, they do not usually occur in the presence of a healthcare professional, which can make diagnosis even more difficult.

- Hallucinations:

Approximately 80% of sufferers experience visual and sometimes auditory hallucinations, often in the early stages of the disease. They are usually realistic and detailed, sometimes discreet in the early stages of the disease as a fleeting sensation.

- Motor symptoms:

Some people with LBD may not experience significant motor problems for several years, while others may suffer from them in the early stages of the disease.

The first symptoms may be very slight and inconsequential, such as a change in handwriting. Akinesia, i.e. difficulty in initiating movement, causes a shuffling gait, blocking, balance problems and falls, a frozen expression and a reduction in the intensity of the voice. All these symptoms can appear at a later stage. If they appear early, the initial diagnosis may be that of Parkinson's disease.

- Sleep disorders:

Behavioral disturbances during REM sleep are frequent. This is when the ill person becomes agitated in their sleep, appearing to be living their dream. They may talk, move violently, fall out of bed or leave it, and continue dreaming in a sort of somnambulistic state.

- Behavioral and mood changes:

Changes in behavior and mood may occur. These usually manifest as symptoms of depression, apathy (*inability to be moved or react*), agitation, anxiety, paranoia or even delirium. Delirium is a false perception of reality. The sufferer may, for example, believe that what they see on television is part of their environment, or that their spouse is having an affair, or that the dead are alive. One type of delusion is specific to LBD. This is the Capgras syndrome (see the note), in which the sufferer believes that a loved one has been replaced by a look-alike who is an impostor. This is a very disturbing phenomenon for which caregivers need to be prepared.

– Diagnosis:

Diagnosis is made by a specialist on the basis of clinical criteria. The main features essential to the diagnosis of LBD are fluctuating cognitive and vigilance disorders, visual hallucinations and physical slowing down with rigidity. Other symptoms are suggestive of LBD, in particular, REM sleep disorders and increased sensitivity to neuroleptics. Brain imaging (CT scan, brain MRI) can detect abnormalities suggestive of Lewy body disease, such as a decrease in the volume of certain areas, and help rule out other possible causes of cognitive and motor disorders. A single-photon emission computed tomography (also called SPECT scan or brain scan) can help establish a diagnosis of LBD. However, no test can diagnose LBD with certainty. The diagnosis can only be confirmed by a post-mortem examination of the brain after death.

A NOTE ON CAPGRAS SYNDROME.–

Capgras' look-alike delusion (sometimes called look-alike illusion, look-alike delusion or Capgras syndrome) is a psychiatric disorder in which the patient, while perfectly able to identify the physiognomy of faces, claims against all odds that the people around them have been replaced by look-alikes who resemble them perfectly. This delusional syndrome was described in 1923 by French psychiatrist Jean Marie Joseph Capgras (1873–1950). A look-alike is a person who bears a perfect resemblance to another.

This condition is more common in patients with neurological or psychiatric disorders, such as schizophrenia (*a severe, chronic mental disorder generally appearing in early adulthood between the ages of around 15 and 30*), traumatic brain injury or dementia. It is more common in patients with neurodegenerative diseases than in those without. It has also been reported to be associated with diabetes, hypothyroidism and migraine attacks. It is more common in women, with a female-to-male ratio of 3:2. Source: https://en.wikipedia.org/wiki/Capgras_delusion, 2021.

4.5.3. *Appendix 3. Parkinson's disease*

4.5.3.1. *Definition*

In 2017, we celebrated the bicentenary of the original description of "shaking palsy" by British physician, geologist and paleontologist *Sir James Parkinson* (1755–1824) in his book *"An essay on the shaking palsy"*, published in 1817. In this essay, Parkinson describes for the first time involuntary shaking movements accompanied by a decrease in muscle strength, leading to a characteristic posture, with the trunk bent forward, and a jerky gait, with the patient suddenly switching from walking to running. His description of the symptoms and progression of the disease remains largely valid today, except for the fact that he claimed there was no cognitive or psychiatric disorder. Indeed, as motor symptoms are clearly visible in the clinic, they were the first and most widely studied, to the detriment of many symptoms today referred to as non-motor. It was not until 1882 (10 years later), when the French neurologist Jean-Martin Charcot (1825–1893) named this agitating paralysis "Parkinson's disease", that light was shed on these non-motor disorders [VAC 18].

Parkinson's disease is a neurodegenerative disorder with a prevalence of over 2% after the age of 65 [AGE 20]. *Prevalence is a measure of the proportion of the population affected by the disease at a given time.* This results in the disappearance of a certain type of neuron: those that produce dopamine almost exclusively. Dopamine (see below) is a neurotransmitter that plays an essential role in controlling movement. Parkinson's disease is characterized by motor disorders, tremors and cognitive problems. The origins of the disease are unknown, but some hypotheses put hereditary factors and environmental factors (or a combination of both) at the root of the condition [BER 21].

Dopamine is a biochemical molecule (Figure 4.9) that enables communication within the nervous system. It is one of the neurotransmitters with a direct influence on behavior. For example, dopamine reinforces daily beneficial actions such as eating a healthy food by inducing the sensation of pleasure.

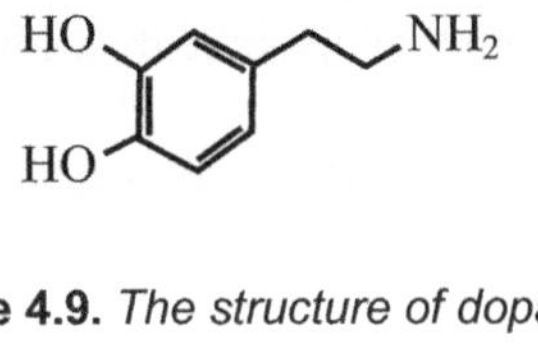

Figure 4.9. *The structure of dopamine*

Individuals suffering from Parkinson's dementia gradually develop rigid, jerky and uncontrollable movements. As it progresses, Parkinson's disease becomes increasingly incapacitating, making everyday tasks such as bathing and dressing difficult, if not impossible. Many of the symptoms of Parkinson's disease are motor symptoms, and concern the control of muscles and movements [MED 18].

The nerve cells affected by Parkinson's disease are located in an area known as the "substantia nigra", in the midbrain. The cells in this area produce dopamine, a chemical messenger (called a neurotransmitter) which not only controls movement, but is also involved in the sensation of pleasure and desire [BAS 17].

4.5.3.2. *Causes*

The death of cells in the substantia nigra results in a lack of dopamine. This leads to an increase in two other chemical messengers: acetylcholine and glutamate. The lack of dopamine thus creates an imbalance in the brain, leading to the onset of Parkinson's symptoms. These symptoms include tremors, muscular rigidity and an inability to perform a number of movements.

Scientists agree that a combination of genetic and environmental factors are involved, although they are not always clearly defined. The current consensus is that *environment* plays a more important role than *heredity*, but that genetic factors predominate when the disease appears before the age of 50. The few environmental factors implicated are as follows [BAS 17]:

– early or prolonged exposure to *chemical pollutants* or *pesticides*, including herbicides and insecticides (e.g. rotenone);

– MPTP, a *drug* sometimes contaminated with heroin, can suddenly cause a severe and irreversible form of Parkinson's disease. Its effect is similar to that of the pesticide rotenone;

– *carbon monoxide* or *manganese* poisoning.

A few definitions are useful to make reading easier:

– *Rotenone* is a natural organic molecule (Figure 4.10) produced by the roots and stems of certain tropical plants. It is a toxic molecule for organisms with mitochondria, as it inhibits complex I of the mitochondrial respiratory chain. It is an ingredient in many pesticides (substances used to control organisms considered harmful) and insecticides (active substances with the ability to kill insects, their larvae and/or eggs). They belong to the pesticide family.

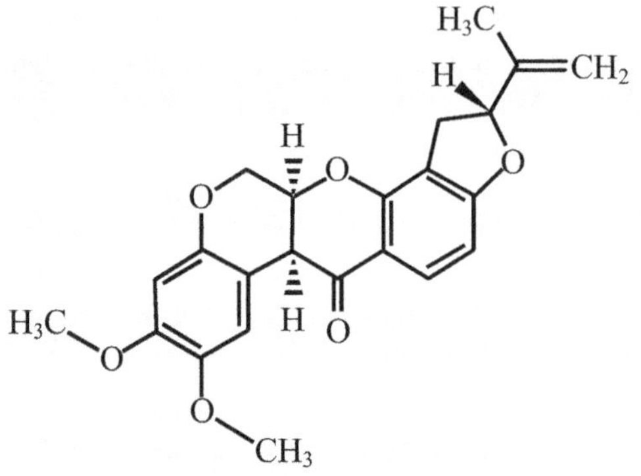

Figure 4.10. *Structure of rotenone*

– *MPTP* (1-*m*ethyl-4-*p*henyl-1,2,3,6-*t*etrahydro*p*yridine) (Figure 4.11) is a neurotoxin that causes the permanent symptoms of Parkinson's disease by destroying certain neurons in the substantia nigra of the brain. While MPTP itself has no *opioid effect*, it is linked to MPPP (1-*m*ethyl-4-*p*henyl-4-*p*ropionoxy*p*iperidine) (Figure 4.11), a synthetic opioid used by drug addicts as a recreational drug.

MPTP, which has effects similar to those of heroin and morphine, can be accidentally produced during the illicit manufacture of MPPP, and it was in this way that its Parkinson's disease-inducing effects were first discovered.

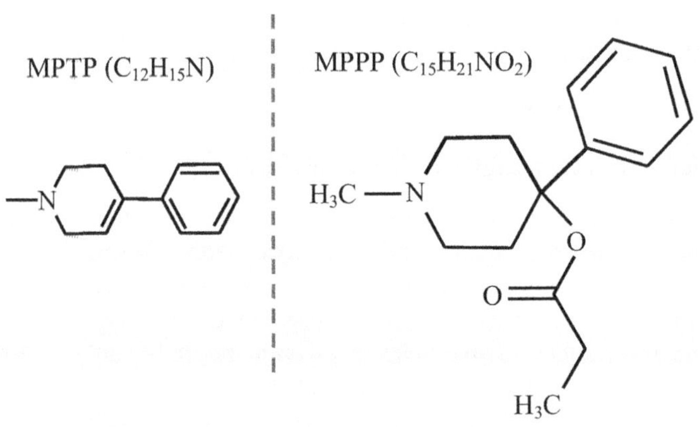

MPTP ($C_{12}H_{15}N$)   MPPP ($C_{15}H_{21}NO_2$)

Figure 4.11. *Structures of MPTP (1-methyl-4-phenyl-1,2,3,6-tetrahydropyridine) and MPPP (1-methyl-4-phenyl-4-propionoxypiperidine)*

Heroin (diamorphine or diacetylmorphine), with formula $C_{21}H_{23}NO_5$ (Figure 4.12), is a molecule used as a drug for its powerful painkilling and euphoric effects. As for morphine, it is a molecule with formula $C_{17}H_{19}NO_3$ (Figure 4.12) used in medicine as an analgesic (drug used to fight pain) and as a drug for its euphoric action.

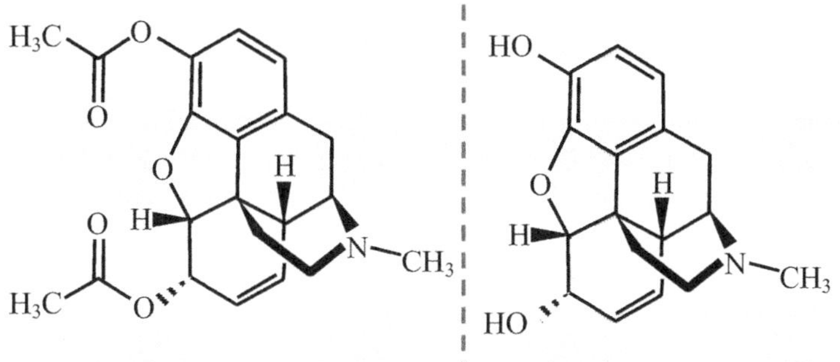

Heroin ($C_{21}H_{23}NO_5$)   Morphine ($C_{17}H_{19}NO_3$)

Figure 4.12. *Structures of heroin and morphine*

VOCABULARY CORNER.–

– The *substantia nigra* is composed of dopaminergic neurons.

– A *psychotropic drug* is a product or chemical substance that acts primarily on the state of the central nervous system.

– Opioids are *psychotropic drugs* that act on the areas of the brain responsible for pain control. Opioids produce an analgesic effect and can induce euphoria. Misuse of opioids can entail health risks.

4.5.3.3. *Symptoms, associated problems*

The four main symptoms of Parkinson's disease are as follows [MED 18]:

– *Tremor* (rhythmic shaking of a limb, the head or the whole body) is the best-known symptom of Parkinson's disease. The tremor often starts with an occasional shake of a finger and then spreads to the whole arm. The tremor may affect only one part or side of the body, particularly in the early stages of the disease. Not everyone with Parkinson's disease has a tremor.

– *Rigidity* (stiffness of limbs and joints). The muscular rigidity associated with Parkinson's disease often begins in the legs and neck. It affects most patients.

Muscles become tense and contracted, and some people may experience pain or stiffness.

– *Bradykinesia or akinesia* (slowness or absence of movement). Bradykinesia is one of the classic symptoms of Parkinson's disease. Over time, Parkinson's patients may arch their backs and walk slowly, dragging their feet. Eventually, they can no longer get into or stay in motion. After a number of years, they may develop significant akinesia and be unable to move at all.

– *Postural instability* (impaired balance and coordination). A person with postural instability may arch their back. They may lean forwards or backwards and fall, leading to injury. People who lean back tend to walk backwards (retropulsion).

Parkinson's disease is often accompanied by treatable problems. These include the following [BAS 17]:

– *Difficulty thinking*. The onset of cognitive problems generally occurs in the later stages of the disease. Such cognitive problems do not respond well to medication.

– *Mood disorders*. People with Parkinson's disease can suffer from depression. With treatment for depression, it is easier to manage the other problems associated with Parkinson's disease. Other disorders such as anxiety or loss of motivation can accompany depression.

– *Swallowing problems*. The person has difficulty swallowing as their condition worsens. Saliva may accumulate in the mouth due to slow swallowing.

– *Sleep disorders*. People with Parkinson's disease often have trouble sleeping. They frequently wake up at night, wake up early or fall asleep during the day.

– *Incontinence*. Parkinson's disease can cause bladder problems, leading to an inability to control urine or difficulty urinating.

– *Constipation*. Many people with the disease become constipated. Constipation is mainly due to a slower digestive tract.

– *Change in blood pressure*, with dizziness or lightheadedness.

– *Impaired sense of smell*. Difficulty identifying or differentiating certain odors.

– *Fatigue*. Many patients experience fatigue, and the cause is not always known.

– *Pain*. Many people with the disease suffer from pain, either in specific areas of the body or throughout the whole body.

– *Sexual dysfunction*. Some sufferers report reduced sexual desire or performance.

4.5.3.4. Worldwide figures for Parkinson's disease

Parkinson's disorders most often appear between the ages of 50 and 70. The average age of onset in Canada and France is 57. At first, symptoms may be confused with normal aging, but as they worsen, the diagnosis becomes more obvious. By the time the first symptoms appear, an estimated 60% to 80% of the nerve cells in the substantia nigra have already been destroyed. By the time symptoms appear, the disease has already progressed an average of 5 to 10 years.

An estimated 6.3 million people worldwide suffer from Parkinson's disease, including 1.2 million in Europe. Parkinson's disease affects around 30,000 people in Belgium. Although this pathology generally appears after the age of 65, around 10% of patients develop Parkinson's disease before the age of 50 [MED 18].

Furthermore, on a worldwide scale, Parkinson's disease is diagnosed in over 300,000 people every year. The number of cases increases with age. It is estimated that by the age of 65, one in 100 people will be affected, and 2 in 100 by the age of 70 and over [BAS 17].

According to global estimates, more than 8.5 million people had Parkinson's disease in 2019. Current estimates suggest that by 2019, Parkinson's disease had resulted in a burden of 5.8 million disability-adjusted life years, an increase of 81% since 2000, and caused 329,000 deaths, an increase of over 100% since 2000 [WHO 22].

References

[AGE 09] AGENCE NATIONALE DE SÉCURITÉ DU MÉDICAMENT ET DES PRODUITS DE SANTÉ, FDG Scan (^{18}F) 260 MBq/mL, solution injectable, available at: http://agence-prd.ansm.sante.fr/php/ecodex/rcp/R0157339.htm, 2009.

[AGE 13a] AGENCE FRANCE-PRESSE, Une personne âgée sur trois meurt avec Alzheimer aux États-Unis, available at: https://www.lapresse.ca/vivre/sante/201303/19/01-4632601-alzheimer-aux-etats-unis.php, 2013.

[AGE 13b]. AGENCE NATIONALE DE SÉCURITÉ DU MÉDICAMENT, Iodure (^{131}I) de sodium pour therapie Mallinckrodt France, gélule, available at: http://agence-prd.ansm.sante.fr/php/ecodex/rcp/R0229270.htm, 2013.

[AGE 20] AGENCE NATIONALE DE LA RECHERCHE (ANR), Les maladies neurodégénératives : le défi des neurosciences, Les cahiers de l'ANR, no. 13, available at: https://anr.fr/fileadmin/documents/2020/ANR_Cahiers_N13_fiches_web%202.pdf, 2020.

[AGE 21] AGENCE INTERNATIONALE DE L'ENERGIE ATOMIQUE, Évaluation en ligne de la lutte contre le cancer au Sénégal par l'AIEA et ses partenaires, available at: https://www.iaea.org/fr/la-lutte-contre-le-cancer-au-senegal, 2021.

[ALA 06] ALAIN C., VANGIONI É., *La nucléosynthèse primordiale : une fenêtre sur la physique des premières minutes de l'Univers*, Éditions Images de la Physique, CNRS, Paris, 2006.

[ALD 48] ALDRICH L.T., NIER A.O., "Argon 40 in potassium minerals", *Physical Review*, vol. 74, pp. 876–877, 1948.

[ALZ 15] ALZHEIMER'S DISEASE INTERNATIONAL, Rapport Mondial Alzheimer 2015 : l'impact global des démences, Report, available at: https://www.alz.co.uk/sites/default/files/pdfs/world-alzheimer-report-2015-summary-sheet-french.pdf, 2015.

[AND 10] ANDERSEN, M.B., STIRLING, C.H., ZIMMERMANN, B. et al., "Precise determination of the open ocean ^{234}U/^{238}U composition", *Geochemistry, Geophysics, Geosystems*, vol. 11, no. 12, 2010.

[AND 18] ANDREI T., Evaluation d'un ylure d'iodonium et d'un dérivé organostannique en vue de la synthèse de la [^{18}F]fluorobenzylamine, Master's Thesis, Université de Liège, Liège, 2018.

[ANK 22] ANKOMAH E., Afrique subsaharienne : la charge du cancer devrait presque doubler dans les 20 prochaines années, available at: https://news.un.org/fr/story/2022/05/1119792, 2022.

[ANS 16] ANSQUER C.F., KRAEBER-BODERE F., "Techniques de médecine nucléaire pour l'imagerie et le traitement des tumeurs neuroendocrines gastro-entéro-pancréatiques", Radiologie et imagerie : abdominale – digestive, doi: 10.1016/S1879-8527(16)67020-X, 2016.

[APP 78] APPLEBY, P.G., OLDFIELD, F., "The calculation of the lead-210 dates assuming aconstant rate of supply of unsupported ^{210}Pb to the sediment", CATENA, vol. 5, pp. 1–8, 1978.

[APP 01] APPLEBY, P.G., "Chronostratigraphic techniques in recent sediments", in LAST W.M., SMOL, J.P. (eds), Tracking Environmental Change Using Lake Sediments, Kluwer Academic Publishers, The Netherlands, 2001.

[ARD 06] ARDISSON V., Evaluation de nouveaux radiopharmaceutiques, PhD Thesis, Université Joseph Fourier-Grenoble 1, Grenoble, 2006.

[ARN 49] ARNOLD J.R., LIBBY W.F., "Age determinations by radiocarbon content: Checks with samples of known age", Science, vol. 110, no. 2869, pp. 678–680, doi: 10.1126/science.110.2869.678, 1949.

[ASS 20] ASSOCIATION DE RECHERCHE SUR LES CANCERS, Épidémiologie des cancers, available at: https://www.arcagy.org/infocancer/le-cancer/epidemiologie-du-cancer.html/, 2020.

[ASS 21] ASSOCIATION POUR LA RECHERCHE SUR LE CANCER, Cancers de la thyroïde : les symptômes et le diagnostic, available at: https://www.fondation-arc.org/cancer/cancer-thyroide/symptomes-diagnostic-cancer, 2021.

[AUT 09] AUTIWA, PHQ953 Nucléosynthèse primordiale stellaire, available at: http://autiwa.free.fr/cours/sem9/PHQ953_Nucleosynthese_primordiale_stellaire.pdf, 2009.

[BAH 19] BAHAIN J.-J., MERCIER N., VALLADAS H., "Datation des sites préhistoriques : quoi de neuf depuis les années 1980 ?", Les nouvelles de l'archéologie, vols 157–158, pp. 107–113, doi: 10.4000/nda.7876, 2019.

[BAI 08] BAILLY S., Le gravitino, candidat à la matière noire et les implications en nucléosynthèse primordiale. PhD Thesis, Université Montpellier II, Montpellier, 2008.

[BAR 56] BARNES J.W., LANG E.J., POTRATZ H.A., "Ratio of ionium to uranium in coral limestone", Science, vol. 123, pp. 175–176, 1956.

[BAR 95] BARO J., SEMPAU J., FERNÁNDEZ-VAREA J.M. et al., "PENELOPE: An algorithm for Monte Carlo simulation of the penetration and energy loss of electrons and positrons in matter", Nuclear Instruments & Methods in Physics Research Section B, vol. 610, pp. 31–36, 1995.

[BAR 15] BARD E., TUNA T., FAGAULT Y. et al., "AixMICADAS, the accelerator mass spectrometer dedicated to ^{14}C recently installed in Aix-en-Provence, France", *Nuclear Instruments and Methods in Physics Research Section B*, vol. 361, pp. 80–86, 2015.

[BAS 17] BASTIANETTO S., La maladie de Parkinson, Institut universitaire de gériatrie de Montréal (IUGM), available at: https://www.passeportsante.net/fr/Maux/Problemes/Fiche.aspx?doc=maladie_parkinson_pm, 2017.

[BEC 10–11] BECK P., Le système solaire, M2 SVT 2010-2011, Laboratoire de Planétologie de Grenoble, available at: https://ipag.osug.fr/~beckp/Research/Teaching_files/Planeto_CAPES_2010.pdf, 2010–2011.

[BEK 06] BEKAERT V., Développement d'un tomographe à émission monophotonique dédié au petit animal, PhD Thesis, Université Louis Pasteur – Strasbourg I, Strasbourg, 2006.

[BÉL 16] BÉLIVEAU R., GRINGAS D., *Les aliments contre le cancer, nouvelle édition revue et augmentée : la prévention du cancer par l'alimentation*, Les Éditions du Trécarré, Montreal, available at: https://ia803003.us.archive.org/13/items/lesalimentscontrelecancer/%5B%20, 2016.

[BEL 18] BELOUALI-DIALLO S., Optimalisation radiopharmaceutique de la scintigraphie gastrique pour un meilleur accès au diagnostic de la gastroparésie, PhD Thesis, Aix-Marseille Université, Marseille, 2018.

[BÊM 19] BÊME D., Les Accidents Vasculaires Cérébraux en chiffres, available at: https://www.doctissimo.fr/html/dossiers/avc/sa_3720_avc_chiffres.htm, 2019.

[BER 21] BERTHON M., Maladie de Parkinson, available at: https://www.deuxiemeavis.fr/pathologie/maladie-de-parkinson, 2021.

[BOI 17] BOISSON T., Qu'est-ce que la limite de Chandrasekhar ?, available at: https://trustmyscience.com/qu-est-ce-que-la-limite-de-chandrasekhar/, 2017.

[BOI 18] BOISSON, T., Quel est l'âge de l'univers et comment est-il déterminé ?, available at: https://trustmyscience.com/age-de-l-univers-comment-est-il-determine/, 2018.

[BOU 03] BOURDON, B., TURNER S., HENDERSON G.M. et al., "Introduction to U-series geochemistry", *Reviews in Mineralogy and Geochemistry*, vol. 52, no. 1, pp. 1–21, 2003.

[BOU 17] BOUTON L., Médicaments radiopharmaceutiques émetteurs de positions : réglementation et production, PhD Thesis, Université de Lille 2, Lille, 2017.

[BOY 22] BOYCE H., HAGGARD D., WITZEL G. et al., "Multiwavelength variability of Sagittarius A* in 2019 July", *The Astrophysical Journal*, vol. 931, no. 1, doi: 10.3837/1538-3357/ac6103, 2022.

[BRE 00] BREISMEISTER J.F., MCNP: A General Monte Carlo N-Particle Transport Code, Technical Report, Version 3C Manual LA-13709-M, 2000.

[BRÉ 06] BRÉMOND A., "À l'aube de la découverte de l'expansion de l'Univers", *Histoire & mesure*, vol. XXI, no. 2, pp. 157–186, available at: https://journals.openedition.org/histoiremesure/1752, 2006.

[BRÉ 08] BRÉMOND A., Vesto Melvin Slipher (1875–1969) et la naissance de l'astrophysique extragalactique, PhD Thesis, Université Claude Bernard – Lyon I, Lyon, 2008.

[BUI 21] BUISSON N., POSTEC A., RICHARDIN P. et al., "Les matériaux du Codex", in BERIOU N., DALARUN J., POIREL D. (eds), Le manuscrit franciscain retrouvé, CNRS Éditions, Paris, 2021.

[BUR 57] BURBIDGE E.M. et al., "Synthesis of the elements in stars", Reviews of Modern Physics, vol. 29, no. 4, doi: 10.1103/RevModPhys.29.537, 1957.

[BUS 13] BUSKULIC D., Notes de cours de PHYS 801. Introduction à la Physique Nucléaire, available at: https://lappweb.in2p3.fr/~buskulic/cours/PHYS801/PHYS801_Physique_Nucleaire.pdf, 2013.

[BUV 09] BUVAT I., Imagerie en médecine nucléaire. Imagerie et modélisation en neurobiologie et cancérologie, Université Paris 11, Orsay, 2009.

[CAS 82] CASSIGNOL C., GILLOT P.Y., "Range and effectiveness of unspiked potassium-argon dating: Experimental groundwork and applications", in ODIN G.S. (ed.), Numerical Dating in Stratigraphy, Wiley, Chichester, 1982.

[CEA 18] COMMISSARIAT À L'ÉNERGIE ATOMIQUE ET AUX ÉNERGIES ALTERNATIVES, Les applications de la radioactivité, available at: https://www.cea.fr/comprendre/Pages/radioactivite.aspx?Type=Chapitre&numero=3, 2018.

[CEN 20] CENTRE HOSPITALIER UNIVERSITAIRE VAUDOIS, Médecine nucléaire, available at: https://www.chuv.ch/fr/medecine-nucleaire, 2020.

[CEN 21] CENTRE D'EXPLORATIONS ISOTOPIQUES, Passer un PET scan, qu'est-ce que cela signifie ? Pourquoi passer un TEP scan ?, available at: https://www.scintigraphie-tep.fr/qu-est-ce-qu-un-tep-scan-ou-petscan.php, 2021.

[CHA 03] CHABAUX, F., RIOTTE, J., DEQUINCEY, O., "UTh-Ra Fractionation during weathering and river transport", in BOURDON B., HENDERSON G.M., LUNDSTROM C.C. et al. (eds), Uranium-series Geochemistry, Mineralogical Society of America, Washington, DC, 2003.

[CHA 10] CHARLASSIER R., Mesure des anisotropies de polarisation du fond diffus cosmologique avec l'interféromètre bolométrique QUBIC, PhD Thesis, Université Paris Diderot-Paris 7, Paris, 2010.

[CHA 15–16] CHARAF C.M., Cours 2. Formation des éléments chimiques, Université Ferhat Abbas Sétif 1, Institut d'Architecture et des Sciences de la Terre, Setif, 2015–2016.

[CHA 21] CHAUVEAU D., Étude du couplage tectonique / érosion / eustatisme sur la morphogenèse des séquences de terrasses de récifs coralliens du Cap Laundi (île de Sumba, Indonésie), PhD Thesis, Université de Bretagne Occidentale, Brest, 2021.

[CHA 22] CHANDRASEKAR T., Cancer de la prostate, available at: https://www.msdmanuals.com/fr/professional/cancer-de-la-prostate, 2022.

[CHE 55] CHERDYNTSEV V.V., CHALOV P.I., KHITRIK M.E., *On Isotopic Composition of Radioelements in Natural Objects and Problems of Geochronology* (in Russian), Trudy III Sessii Komissi Oprend, Absol Vozrasta Izd.Akad, Nauk, SSSR, Moscow, 1955.

[CHE 86] CHEN J.H., EDWARDS R.L., WASSERBURG G.J., "^{238}U, ^{234}U and ^{232}Th in seawater", *Earth and Planetary Science Letters*, vol. 80, nos 3–4, pp. 241–251, 1986.

[CHE 00] CHENG H., EDWARDS R.L., HOFF J. et al., "The halflives of uranium-234 and thorium-230", *Chemical Geology*, vol. 169, pp. 17–33, 2000.

[CHE 13] CHENG H., EDWARDS R.L., SHEN C.C. et al., "Improvements in ^{230}Th dating, ^{230}Th and ^{234}U half-life values, and U-Th isotopic measurements by multi-collector inductively coupled plasma mass spectrometry", *Earth and Planetary Science Letters*, vols 371–372, pp. 82–91, 2013.

[CHO 20] CHOI S.K., HASSELFIELD M., HO S.-P.P. et al., "The Atacama Cosmology Telescope: A measurement of the cosmic microwave background power spectra at 98 and 150 GHz", *JCAP12*, vol. 35, doi: 10.1088/1375-7516/2020/12/035, 2020.

[CIM 79] CIMETIÈRE C., LÉGER J., MUXART R. et al., Utilisation de cyclotrons pour la préparation de radionucléides à usage médical, in *Séminaire sur les accélérateurs de particules : utilisation – radioprotection*, Saclay, CBA – CONF 3631, 13–19 May 1979.

[COB 03] COBB K.M., CHARLES C.D., CHENG H. et al., "U/Th-dating living and young fossil corals from the central tropical Pacific", *Earth and Planetary Science Letters*, vol. 210, nos 1–2, pp. 91–103, 2003.

[COL 19] COLEIRO A., L'Univers de haute énergie, Lecture notes, Laboratoire APC, Université Paris Diderot, Paris, available at: http://html5.ens-lyon.fr/Acces/Astronomie/20190820/20190820_LUniversDeHauteEnergie_1.pdf, 2019.

[COM 12] COMMISSARIAT À L'ENERGIE ATOMIQUE, Mesurer l'âge de l'univers, available at: http://bonnetbidaud.free.fr/pedagogie/hubble_law/web_database/Age_de_univers.pdf, 2012.

[COM 15] COMMISSARIAT À L'ENERGIE ATOMIQUE, Nucléosynthèse primordiale, available at: https://www.cea.fr/comprendre/Pages/matiere-univers/astrophysique-nucleaire.aspx?Type=Chapitre&numero=3, 2015.

[COM 16] COMMISSION CANADIENNE DE SÛRETÉ NUCLÉAIRE, Imagerie médicale et radiothérapie, available at: https://www.cnsc-ccsn.gc.ca/fra/resources/infographics/mir/index.cfm, 2016.

[COM 17] COMMISSARIAT À L'ENERGIE ATOMIQUE, Les étoiles, available at: https://www.cea.fr/comprendre/Pages/matiere-univers/essentiel-sur-les-etoiles.aspx, 2017.

[COM 22] COMMISSION CANADIENNE DE SÛRETÉ NUCLÉAIRE, Effets de l'accident de Tchernobyl sur la santé, available at: https://nuclearsafety.gc.ca/fra/resources/health/health-effects-chernobyl-accident.cfm, 2022.

[COR 08] CORVOL P., "VEGF, anti-VEGF et pathologies", *Bulletin de l'Academie nationale de medecine*, vol. 192, no. 2, pp. 289–302, 2008.

[COU 18] Coulon J.-P., Production et utilisation des radioéléments, in *6ème séminaire : médecine et nucléaire*, Cadarache, CEA, available at: https://docplayer.fr/79889717-Production-et-utilisation-des-radioelements.html, 2018.

[CRO 83] Cross T.S., Cross B.W., "U, Sr, and Mg in Holocene and Pleistocene corals *A. palmata* and *M. annularis*", *Journal of Sedimentary Research*, vol. 53, no. 2, pp. 587–594, 1983.

[CSM 21] Centre Du Sein De Montreal VM-Med, Médecine nucléaire et TEP/CT, available at: https://www.vmmed.com/fr/medecine-nucleaire-et-tepct/tepct-en-oncologie/, 2021.

[DAI 13] Daigle S.M., Low energy proton capture study of the ^{13}N(p, γ)^{15}O reaction, PhD Thesis, University of North Carolina, Chapel Hill, doi: 10.17615/nhbv-w333, 2013.

[DAL 69] Dalrymple G.B., Lanphere M.A., *Potassium-Argon Dating: Principles, Techniques and Applications to Geochronology*, 1st ed., W.H. Freeman, New York, 1969.

[DAM 89] Damon P.E., Donahue D.J., Gore B.H. et al., "Radiocarbon dating of the Shroud of Turin", *Nature*, vol. 337, pp. 611–615, doi: 10.1038/337611a0, 1989.

[DEB 08] De Beco V., Le Bars D., Scherrmann J.-.M., "Le fluor 18 en radiopharmacie. Fluorine-18 in radiopharmacy", *Annales pharmaceutiques françaises*, vol. 66, no. 1, pp. 60–65, 2008.

[DEF 08] De Finance L., "La datation des objets : quelques exemples concrets", *In Situ*, vol. 9, doi: 10.4000/insitu.3513, 2008.

[DEG 17] Degrelle D., Caractérisation numérique de la technique de spectrométrie gamma par simulation Monte-Carlo. Application à la datation d'échantillons environnementaux, PhD Thesis, Université Bourgogne Franche Comté, Besançon, 2017.

[DEL 02] Delanghe D., Bard E., Hamelin B., "New TIMS constraints on the uranium-238 and uranium-234 in seawaters from the main ocean basins and the Mediterranean Sea", *Marine Chemistry*, vol. 80, pp. 79–93, 2002.

[DEL 17] Delbecq D., La foudre déclenche des réactions… nucléaires, available at: https://www.lemonde.fr/sciences/la-foudre-declenche-des-reactions-nucleaires, 2017.

[DEL 23] De La Vaissière C. et al., Retombées essais nucléaires, available at: https://laradioactivite.com/energie_nucleaire/retombees_essais_nucleaires, 2023.

[DES 00] Desuzinges D., *Les radiopharmaceutiques et la radiopharmacie : aspects réglementaires et techniques*, École Normale de la Santé Publique, Rennes, 2000.

[DES 13] Descôteaux C., Développement de nouveaux composés anticancéreux pour le traitement des cancers féminins, PhD Thesis, Université du Québec à Trois Rivières (UQTR), Quebec, 2013.

[DEV 53] De Vries H., Barendsen G.W., "Radiocarbon dating by a proportional counter filled with carbon dioxide", *Physica (Amsterdam)*, vol. 19, p. 987, 1953.

[DEV 18] Deviese T., Comeskey D., McCullagh, J. et al., "New protocol for compound-specific radiocarbon analysis of archaeological bones", *Rapid Communications in Mass Spectrometry*, vol. 32, no. 5, pp. 373–379, doi: 10.1002/rcm.8047, 2018.

[DIE 22] Dieng M.D., Cancer de la prostate, radioscopie d'une tumeur fatale aux hommes, available at: https://www.seneplus.com/sante/radioscopie-dune-tumeur-fatale-aux-hommes, 2022.

[DIO 74] Diop C.A., *Physique nucléaire et chronologie absolue*, IFAN-Dakar et les Nouvelles Éditions Africaines, Dakar, 1974.

[DOM 18] Dominguez E., La démence à corps de Lewy, PhD Thesis, Université de Picardie Jules Verne, Amiens, 2018.

[DON 02] Donahue D.J., Olin J.S., Harbottle G., "Determination of the radiocarbon age of parchment of the Vinland Map", *Radiocarbon*, vol. 44, pp. 45–52, doi: 10.1017/S0033822200064651, 2002.

[DUC 15] Duchemin C., Étude de voies alternatives pour la production de radionucléides innovants pour les applications médicales, PhD Thesis, Université de Nantes, Nantes, 2015.

[DUP 01] Dupont Y., Distance, mouvement, masse et rayon des étoiles, available at: https://craq-astro.ca/phy1971/chap_pdf/chap15b.pdf, 2001.

[DUR 23] Duran M., Lerozier C., Comment calculer son IMC (Indice de Masse Corporelle) ?, available at: https://www.santemagazine.fr/minceur/imc-indice-de-masse-corporelle-267579, 2023.

[ÉCO 13] École de Technologie Supérieure, GTS503_C3-Médecine Nucléaire, available at: https://cours.etsmtl.ca/gts503/Cours/GTS503_C3-ImagerieNucleaire.pdf, 2013.

[EDW 86] Edwards R.L., Chen J.H., Wasserburg G.J., "238U-234U-230Th-232Th systematics and the precise measurement of time over the past 500,000 years", *Earth and Planetary Science Letters*, vol. 81, pp. 175–192, 1986.

[EDW 87] Edwards, R.L., Chen, J.H., Ku, T.-L. et al., "Precise timing of the last interglacial period from mass spectrometric determination of thorium-230 in corals", *Science*, vol. 236, pp. 1547–1553, 1987.

[EDW 03] Edwards R.L., Gallup C.D., Cheng H., "Uranium-series dating of marine and lacustrine carbonates", in Bourdon B., Henderson G.M., Lundstrom C.C. et al. (eds), *Uranium-series Geochemistry*, Mineralogical Society of America, Washington, DC, 2003.

[ESS 97] Esser R.P., McIntosh W.C., Heizler M.T. et al., "Excess argon in melt inclusions in zero-age anorthoclase feldspar from Mt Erebus, Antarctica, as revealed by the ^{40}Ar/^{39}Ar method", *Geochimical et Cosmochimica Acta*, vol. 61, no. 18, pp. 3789–3801, 1997.

[ESS 19] Esslinger O., La nucléosynthèse primordiale, available at: https://www.astronomes.com/le-big-bang/nucleosynthese-primordiale, 2019.

[FAR 17] Farrell J., The tangled history of big bang science, available at: https://nautil.us/blog/the-tangled-history-of-big-bang-science, 2017.

[FER 04] FERNANDO M.S., INCE P.G., "Vascular pathologies and cognition in a population based cohort of elderly people", *Journal of the Neurological Sciences*, vol. 226, pp. 13–17, doi: 10.1016/j.jns.2004.09.004, 2004.

[FEY 19] FEYROUZ L., MEBARKA L., Étude de la réalisation d'un détecteur de rayons gamma, Master's Thesis, Université Abou Bakr Belkaïd de Tlemcen, Algeria, 2019.

[FON 04] FONTUGNE M., "Les derniers progrès du calibrage des âges radiocarbone permettent-ils une révision des chronologies entre 25 et 50.000 ans B.P. ?" *Quaternaire*, vol. 15, no. 3, pp. 245–252, doi: 10.3406/quate.2004.1771, 2004.

[FON 20] FONDATION CONTRE LE CANCER, PET-scan. Tomographie par émission de positrons, Chaussée de Louvain 379, 1030 Brussels, available at: https://www.cancer.be/les-cancers/le-pet-scan-imagerie-m-tabolique, 2020.

[FRA 09] FRAJNDLICH R., "IEA-R1 research reactor: Operational life extension and considerations regarding future decommissioning", *International Nuclear Atlantic Conference - INAC 2009 Rio de Janeiro*, 2009.

[FRA 14] FRANÇOIS A., Elaboration de complexes hétérobinucléaires par approche chimie Click pour une application en imagerie bimodale, PhD Thesis, Université de Toulouse, 2014.

[FRE 01] FREEDMAN W.L., MADORE B.F., GIBSON B.K. et al., "Final results from the Hubble Space Telescope key project to measure the Hubble constant", *Astrophysical Journal*, vol. 553, pp. 37–72, doi: 10.1086/320638, 2001.

[FRE 13] FREER M., FAYNBO H., "The Hoyle state in 12C", *Progress in Particle and Nuclear Physics*, vol. 78, pp. 1–23, doi: 10.1016/j.ppnp.2013.06.001, 2013.

[FRE 22] FREREBEAU N., LEBRUN B., "Calibrer des âges radiocarbone avec R". *Programming Historian*, doi: 10.46430/phfr0016, 2022.

[FRI 04] FRIEDRICH M., REMMELE S., KROMER B. et al., "The 12,460 year Hohenheim oak and pine tree ring chronology from central Europe – A unique annual record for radiocarbon calibration and paleoenvironment reconstructions", *Radiocarbon*, vol. 46, no. 3, pp. 1111–1122, 2004.

[FRU 18] FRUET G., Structure des ions lourds et nucléosynthèse dans les étoiles massives : la réaction $^{12}C + {}^{12}C$, PhD Thesis, Université de Strasbourg, Strasbourg, 2018.

[FUS 72] FUSCO M.A., PEEK N.F., JUNGERMANN, F.W. et al. "Cyclotron production of carrier-free ^{123}I using the reaction ^{127}I (p, 5n) ^{123}Xe", *The Journal of Nuclear Medicine*, vol. 13, no. 10, pp. 729–732.

[GAR 75] GARNER E.L., MURPHY T.J., GRAMLICH J.W. et al., "Absolute isotopic abundance ratios and the atomic weight of a reference sample of potassium", *Journal of Research of the National Bureau of Standards*, vol. 79A, pp. 713–725, 1975.

[GAS 82] GASCOYNE M., "Geochemistry of the actinides and their daughters", in IVANOVITCH M., HARMON R.S. (eds), *Uranium Series Disequilibrium: Applications to Environmental Problems*, Clarendon Press, Oxford, 1982.

[GAS 18] GASAGHRANDE K., 68Ga-PSMA-11, nouveau traceur TEP pour l'imagerie di cancer de la prostate : synthèse, contrôles qualité et dossier d'autorisation. PhD Thesis, Université Toulouse III, Toulouse, 2018.

[GIL 82] GILLOP P.-Y., CHIESA S., PASQUARE G. et al., "33,000 years K-Ar dating of the volcano-tectonic horst of the Isle of Ischia, Gulf of Naples". *Nature*, vol. 299, no. 5880, pp. 242–244, 1982.

[GIL 86] GILLOT P.-Y., CORNETTE Y., "The Cassignol technique for potassium-argon dating, precision and accuracy: Examples from late Pleistocene to recent volcanics from southern Italy", *Chemical Geology*, vol. 59, pp. 205–222, 1986.

[GIL 06] GILLOT P.-Y., HILDENBRAND A., LEFÈVRE J.-C. et al, "The K/Ar dating method: Principle, analytical techniques, and application to Holocene volcanic eruptions in Southern Italy", *Acta Vulcanologica*, vol. 18, pp. 55–66, 2006.

[GIO 19] GIORGETTA J., PET-Scan : pourquoi en faire, quels effets secondaires après ?, available at: https://sante.journaldesfemmes.fr/fiches-anatomie-et-examens/, 2019.

[GOU 75] GOURINARD Y., "Méthode Potassium-Argon et chronologie quaternaire", *Bulletin de l'Association française pour l'étude du quaternaire*, vol. 12, no. 2, pp. 83–89, doi: 10.3406/quate.1975.2085, 1975.

[GUI 98] GUILLOU H., CARRACEDO J.C., DAY S.J., "Dating of the Upper Pleistocene-Holocene volcanic activity of La Palma using the unspiked K-Ar technique", *Journal of Volcanology and Geothermal Research*, vol. 86, pp. 137–149, 1998.

[GUI 17] GUILLOU H., SCAO V., NOMADE S. "De la justesse des âges K-Ar : exemple de la datation de deux dômes trachytiques du Gölcük (Turquie)", *Quaternaire*, vol. 28, no. 2, pp. 141–148, doi: 10.4000/quaternaire.7975, 2017

[GUI 18] GUIBERT P., "Dater, une histoire qui date", *ArcheoSciences, revue d'archéométrie*, vol. 42, no. 1, pp. 85–101, doi: 10.4000/archeosciences.5390, 2018.

[HAB 13] HABERT M.O., KAS A., "Aspects techniques et pratiques, et indications de la tomographie d'émission monophotonique cérébrale", *EMC-Neurologie*, vol. 10, no. 3, doi: 10.1016/S0236-0378(13)52780-1, 2013.

[HAN 16] HANANE A., Conception et réalisation d'une banque de données médicale, Master's Thesis, Université Abou Bakr Belkaïd de Tlemcen, Algeria, 2015–2016.

[HAR 81] HARRISON T.M., MCDOUGALL I., "Excess ^{40}Ar in metamorphic rocks from Broken Hill, New South Wales: Implications for ^{40}Ar/^{39}Ar age spectra and the thermal history of the region", *Earth and Planetary Science Letters*, vol. 55, no. 1, pp. 123–149, 1981.

[HER 19] HERBERT A., Mise au point de nouvelles techniques de radio-iodation et application au radiomarquage de molécules d'intérêt, PhD Thesis, Université Caen Normandie, Caen, 2019.

[HER 20] HERTZ B., BHARADWAJ P., GREENSPAN B., "A historic review of the discovery of the medical uses of radioiodine", *Pakistan Journal of Nuclear Medicine*, vol. 10, no. 1, pp. 1–4, doi: 10.24911/PJNMed.175-1582813482, 2020.

[HIG 21] HIGHAM T., BASELL L., JACOBI R. et al., "Testing models for the beginnings of the Aurignacian and the advent of figurative art and music: The radiocarbon chronology of Geißenklösterle", *Journal of Human Evolution*, vol. 62, no. 6, pp. 664–676, doi: 10.1016/j.jhevol.2012.03.003, 2021.

[HOL 11] HOLDEN N.E., BONARDI M.L., DE BIÈVRE P. et al., "Common definition and convention on the use of the year as a derived unit of time (IUPAC Recommendations 2011)", *Pure and Applied Chemistry*, vol. 83, no. 5, pp. 1159–1162, doi: 10.18814/epiiugs/2011/v34i1/006, 2011.

[HUB 29] HUBBLE E., "A relation between distance and radial velocity among extra-galactic nebulae", *Proceedings of the National Academy of Sciences*, vol. 15, no. 3, pp. 168–173, 1929.

[HUB 04] HUBERT P., HUBERT F., RAFFESTIN-TORT V., "La datation des vins : une application des mesures des très faibles radioactivités", *Bulletin de l'union des physiciens*, vol. 98, no. 862, pp. 381–395, 2004.

[IMB 05] IMBRIANI G. et al., "S-factor of ^{13}N (p, γ)^{15}O at astrophysical energies", *European Physical Journal A*, vol. 25, pp. 355–366, doi: 10.1130/epja/i2005-10138-7, 2005.

[INS 10] INSTITUT NATIONAL DU CANCER, Les traitements du cancer de la prostate, available at: https://www.e-cancer.fr, 2010.

[INS 19] INSTITUT DE RADIOPROTECTION ET DE SÛRETÉ NUCLÉAIRE (IRSN), Les principes de la radiothérapie, available at: https://www.irsn.fr/savoir-comprendre/sante/principes-radiotherapie#, 2019.

[INS 21] INSTITUT NATIONAL DU CANCER, Données globales d'épidémiologie des cancers, available at: https://www.e-cancer.fr/Professionnels-de-sante/Epidemiologie-des-cancers/Donnees-globales, 2021.

[INS 22] INSTITUT DU CERVEAU, La maladie d'Alzheimer : causes, mécanismes biologiques, symptômes et diagnostic, traitements, available at: institutducerveau-icm.org, 2022.

[JAF 71] JAFFEY A.H., FLYNN K.F., GLENDENIN L.E. et al. "Precision measurement of half-lives and specific activities of ^{235}U and ^{238}U", *Physical Review C*, vol. 4, no. 5, 1889–1906, 1971.

[JAN 15] JANEVIK E., "Radiopharmaceuticals in neurological and psychiatric disorders", in *International Conference on Clinical PET-CT and Molecular Imaging (IPET 2015): PET-CT in the Era of Multimodality Imaging and Image-Guided Therapy*, Vienna, 5–9 October 2015.

[JEN 20] JENNIFER L., Administration des medicaments, Skaggs School of Pharmacy and Pharmaceutical Sciences, University of California, San Diego, available at: https://www.msdmanuals.com/fr/accueil/m%C3%A9dicaments/, 2020.

[JOY 13] JOYARD Y., Synthèse de nouveaux radiomarqueurs potentiels de l'hypoxie tumorale : développement d'une nouvelle méthodologie de fluoration nucléophile et son application vers la synthèse du 2-[18F]Fluoro-2-désoxy-D-glucose, PhD Thesis, INSA de Rouen, 2013.

[JUL 03] JULL A.J.T., "Radiocarbon", in *18th Conference*, Wellington, vol. 36, available at: http://www.radiocarbon.org/, 2003.

[KEL 02] KELLY S., "Excess argon in K-Ar and Ar-Ar geochronology", *Chemical Geology*, vol. 188, nos 1–2, pp. 1–22, doi: 10.1016/S0009-2541(02)00064-5, 2002.

[KIR 13] KIRSEBOM O.S., "^{12}C and the triple-α reaction rate", *Journal of Physics: Conference Series*, vol. 336, p. 012072, doi: 10.1088/1732-6596/336/1/012072, 2013.

[LAC 11] LACOEUILLE F., Conception, développement et évaluation d'un nouveau radiopharmaceutique pour la scintigraphie de la perfusion pulmonaire, PhD Thesis, Université d'Angers, Angers, 2011.

[LAL 67] LAL D., PETERS B., "Cosmic ray produced radioactivity on the Earth", in SITTLE K. (ed.), *Kosmische Strahlung II / Cosmic Rays II*, Springer-Verlag, Berlin, pp. 551–612, doi: 10.1007/978-3-642- 46079-1_7, 1967.

[LAL 79] LALOU C., HOANG C.-T., "Les méthodes de datation par les descendants de l'Uranium", *Bulletin de l'Association française pour l'étude du quaternaire*, vol. 16, nos 1–2, pp. 3–13, doi: 10.3306/quate.1979.1331, 1979.

[LAP 12] LAPORTE B., Analyse de concordance entre IRM de perfusion de stress et la scintigraphie myocardique à partir des 3 premières années d'expérience nantaise, PhD Thesis, Université de Nantes, Nantes, 2012.

[LAS 17–21] LA SOCIÉTÉ ALZHEIMER DE QUÉBEC, La maladie d'Alzheimer en chiffres, available at: https://www.societealzheimerdequebec.com/comprendre-la-maladie/statistiques/ 2017, 2021.

[LAU 94] LAUGHLIN A.W., POTHS J., HEALEY H.A. et al., "Dating of quaternary basalts using the ^{3}He and ^{14}C methods with implications for excess ^{40}Ar", *Geology*, vol. 22, pp. 135–138, 1994.

[LEE 06] LEE, J.-Y., MARTI, K., SEVERINGHAUS, J.P. et al., "A redetermination of the isotopic abundances of atmospheric Ar", *Geochimica et Cosmochimica Acta*, vol. 70, no. 17, pp. 4507–4512, 2006.

[LEF 94] LEFÈVRE J.-CL., GILLOT P.-Y., "Datation Potassium-Argon de roches volcaniques du Pléistocènes supérieur et de l'Holocène. Exemple de l'Italie du Sud ; application à l'archéologie", *Bulletin de la Société Préhistorique Française*, vol. 91, no. 2, pp. 145–148, 1994.

[LEH 08] LEHAUTE, F., Biochimie des contaminants organiques HAP, PCB et pesticides organochlores dans les sédiments de l'étang de Thau. PhD Thesis, Université Pierre et Marie Curie, Paris, 2008.

[LEH 11] LE HÔ A.-S., BOREL T., LAVIER C. et al., "Examens, analyses et datation de trois portraits en cire d'Henri IV – À la recherche des techniques et d'une chronologie", in *Conference Proceedings*, Institut de France, Fondation de Chantilly, June 2010, pp. 39–52, 2011.

[LEM 27] LEMAÎTRE G., "Un Univers homogène de masse constante et de rayon croissant rendant compte de la vitesse radiale des nébuleuses extragalactiques", *Annales de la société scientifique de Bruxelles*, vol. A37, pp. 39–59, 1927.

[LIB 49] LIBBY W.F., ANDERSON E.C., ARNOLD J.R., "Age determination by radiocarbon content: World-wide assay of natural radiocarbon", *Science*, vol. 109, no. 2827, pp. 227–228, 1949.

[LOM 20] LOMBRISER L., Le mystère de l'expansion de l'Univers trouve une solution, Université de Genève, doi: 10.1016/j.physletb.2020.135303, 2020.

[LUD 03] LUDWIG K.R., "Mathematical-statistical treatment of data and errors for ^{230}Th/U geochronology", *Reviews in Mineralogy and Geochemistry*, vol. 52, no. 1, pp. 631–656, doi: 10.2113/0520631, 2003.

[LUM 97] LUMINET J.-P., *L'invention du Big Bang*. Le Seuil, Paris, available at: https://luth.obspm.fr/~luminet/Books/FL.html, 1997.

[MAB 08] MABIT L., BENMANSOUR M., WALLING D.E., "Comparative advantages and limitations of the fallout radionuclides ^{137}Cs, ^{210}Pbex and ^{7}Be for assessing soil erosion and sedimentation", *Journal of Environmental Radioactivity*, vol. 99, pp. 1799–1807, 2008.

[MAG 21] MAGNARD B, SEYNAT M., Réaliser une chromatographie sur couche mince, Maxicours, available at: https://www.maxicours.com/se/cours/realiser-une-chromatographie-sur-couche-mince/, 2021.

[MAI 19] MAIRESSE F., "Géopolitique du musée : les enjeux de la fréquentation", *Politique et Sociétés*, vol. 38, no. 3, pp. 103–127, doi: 10.7202/1064732ar, 2019.

[MAJ 14] MAJDOULINE EL.H., Les radiopharmaceutiques de la production à l'injection aux patients : l'exemple du ^{18}FDG, PhD Thesis, Université Mohammed V – Souissi, Rabat, 2014.

[MAL 13]. MALLET E., Comment mieux comprendre le métabolisme de la vitamine D ?, available at: https://www.realites-cardiologiques.com/wp-content/uploads/sites/2/2013/11/Mallet_metabolismevitD.pdf, 2013.

[MAL 23] MALVEZZI M., SANTUCCI C., BOFFETTA P. et al., "European cancer mortality predictions for the year 2023 with focus on lung cancer", *Annals of Oncology*, doi: 10.1016/j.annonc.2023.01.010, 2023.

[MAN 20] MANGEANT R., Radiotraceurs des récepteurs sérotoninergiques utilisés chez l'homme, PhD thesis, Université Caen Normandie, Caen, 2020.

[MAR 10] MARTIN J.-P., Conférence : LE BIG BANG POUR LES NULS, available at: https://www.planetastronomy.com/special/2011-special/13nov10/jpm-bb.htm, 2010.

[MAR 12] MARTA M., The ^{13}N(p,γ)^{15}O reaction studied at low and high beam energy, available at: https://www.hzdr.de/publications/PublDoc-5915.pdf, 2012.

[MAS 99] MASARIK J., BEER J., "Simulation of particle fluxes and cosmogenic nuclide production in the Earth's atmosphere", *Journal of Geophysical Research*, vol. 104, pp. 12099–12111, doi: 10.1029/1998JD200091, 1999.

[MCD 14] MCDOUGALL I., "K/Ar and $^{40}Ar/^{39}Ar$ isotopic dating techniques as applied to young volcanic rocks, particularly those associated with hominin localities", in Davis A.M. (ed.), *Treatise on Geochemistry*, 2nd ed., Elsevier, Amsterdam, doi: 10.1016/B978-0-08-095975-7.01201-8, 2014.

[MED 18] MEDTRONIC BELGIUM, À propos de la maladie de Parkinson, available at: https://www.medtronic.com/be-fr/patients/pathologies/maladie-de-parkinson.html, 2018.

[MIN 95] MIN G.R., EDWARDS R.L., TAYLOR F.W. et al., "Annual cycles of UCa in coral skeletons and UCa thermometry", *Geochimica et Cosmochimica Acta*, vol. 59, no. 10, pp. 2025–2042, 1995.

[MON 18] MONTEMAGNO C., Développement de radiotraceurs pour l'imagerie phénotypique des cancers du sein métastatiques, PhD Thesis, Université Grenoble Alpes, Grenoble, 2018.

[MUL 77] MULLER R.A., "Radio-isotope dating with a cyclotron", *Science*, vol. 196, no. 4289, pp. 489–494, 1977.

[MYE 79] MYERS W.G., "Georg Charles de Hevesy: The father of nuclear medicine", *Journal of Nuclear Medicine*, vol. 20, pp. 590–593, 1979.

[NDA 23] NDAR INFO, Cancer : 8 mille personnes tuées par an au Sénégal, available at: https://www.ndarinfo.com/Cancer-8-mille-personnes-tuees-par-an-au-Senegal_a35278.html, 2023.

[NDE 11] NDEYE M., SÉNE M., DIALLO A.O., "IFAN radiocarbon laboratory measurement I", *Radiocarbon*, vol. 53, no. 1, pp. 167–17, 2011.

[NDE 12] NDEYE M., SENE M., DIALLO A.O., "Datation par la méthode du radiocarbone avec un compteur à scintillation liquide muni de l'option "Super Low Level"". *Bulletin IFAN Ch. A. Diop, TL III, série. A*, no. 1, pp. 123–139, Dakar, 2012.

[NGU 23] N'GUESSAN K.J.-F., SAKHO I., KA K. et al., "Secondary cancer risk assessment after high-risk and intermediate-risk prostate cancer radiotherapy in Senegal", *World Journal of Advanced Research and Reviews*, vol. 17, pp. 230–238, 2023.

[NIC 10] NICOL S., Étude et construction d'un tomographe TEP/TDM pour petits animaux, combinant modules phoswich à scintillateurs et détecteur à pixels hybrides, PhD Thesis, Université de la Méditerranée – Aix-Marseille II, Marseille, 2010.

[NIE 50] NIER, A.O., "A redetermination of the relative abundances of the isotopes of carbon, nitrogen, oxygen, argon and potassium", *Physical Reviews*, vol. 77, pp. 789–793, 1950.

[NOM 17] NOMADE S., "Recommandation sur l'utilisation des unités de temps en sciences de la terre", *Quaternaire*, vol. 28, no. 2, pp. 137–139, doi: 10.4000/quaternaire.7972, 2017.

[OPE 07] OPECST, La médecine nucléaire, available at: https://www.senat.fr/opecst/note/2007_1_medecine_nucleaire_3p.pdf, 2007.

[PAL 07] PALLARDY G., PALLARDY M.-J., "Histoire abrégée du radiodiagnostic et de l'imagerie médicale", *Histoire des sciences médicales*, vol. XLI, no. 1, pp. 34–40, available at: http://www.biusante.parisdescartes.fr/sfhm/hsm/HSMx2007x041x001/HSMx20 07x041x001x0034.pdf, 2007.

[PAP 09] PAPASTEFANOU, C., "Beryllium-7 aerosols in ambient air. aerosol and air", *Quality Research*, vol. 9, no. 2, pp. 187–197, 2009.

[PAT 12] PATUREL G., L'effet Doppler Fizeau, available at: http://clea-astro.eu/lunap/DopplerFizeau, 2012.

[PAY 08] PAYOUK P., ALONSO M., ESQUERRÉ J.-P. et al. "Les "nouveaux" radiopharmaceutiques", *Médecine Nucléaire*, vol. 32, no. 8, pp. 356–361, doi: 10.1016/j.mednuc.2008.06.007, 2008.

[PHI 20] PHILIPPON J., "Datation et mise en valeur de l'icône dite de "Notre-Dame de Grâce" dans la cathédrale de Cambrai", *Monumental*, vol. 1, pp. 16–17, 2020.

[POI 19] POIRIER-QUINOT M., *L'Imagerie Médicale, quand la physique rencontre la médecine*, École Normale Supérieure Paris-Saclay, Paris, 2019.

[POR 03] PORCELLI D., SWARZENSKI P.W., "The behaviour of U- and Th-nuclides in groundwater", *Geochemistry*, vol. 52, no. 1, pp. 317–361, 2003.

[POT 55] POTRATZ H.A., BARNES J.W., LANG E.J. et al., *A Radio-Chemical Procedure for Thorium and Its Application to the Determination of Ionium in Coral Limestone*, Los Alamos Scientific Laboratory of the University of California, Los Alamos, 1955.

[QAI 11] QAIM S.M., TARKANYI F., CAPOTE R., Nuclear data for the production of therapeutic radionuclides, IAEA, Vienna, available at: https://www-pub.iaea.org/MTCD/publications/PDF/trs373_web.pdf, 2011.

[QUI 13] QUILES A., AUBOURG E., BERTHIER B. et al., "Bayesian modelling of an absolute chronology for Egypt's 18th Dynasty by strophysical and radiocarbon methods", *Journal of Archaeological Science*, vol. 40, no. 1, pp. 423–432, 2013.

[RAD 21] RADIOLOGICAL SOCIETY OF NORTH AMERICA, Tomographie par émission de positons – Tomodensitométrie (TEP/CT), available at: https://www.radiologyinfo.org/en/info/pet, 2021.

[RAT 19] RATHMANN, S.M., AHMAD, Z., SLIKBOER, S. et al., "The radiopharmaceutical chemistry of technetium-99m", *Radiopharmaceutical Chemistry*, pp. 311–333, doi: 10.1007/978-3-319-98937-1_18, 2019.

[REI 04] REIMER P.J., BAILLIE M.G.L, BARD E. et al., "IntCal04 terrestrial radiocarbon age calibration, 0-26 cal ka BP", *Radiocarbon*, vol. 46, pp. 1029–1058, 2004.

[REI 09] REIMER P.J., BAILLIE M.G.L., BARD E. et al., "IntCal09 and Marine09 radiocarbon age calibration curves, 0–50,000 years cal BP", *Radiocarbon*, vol. 51, no. 4, pp. 1111–1150, doi: 10.1017/S0033822200034202, 2009.

[REI 13] REIMER P.J., BAILLIE M.G.L., BARD E. et al., "IntCal13 and Marine13 radiocarbon age calibration curves 0–50,000 years cal BP", *Radiocarbon*, vol. 55, no. 4, pp. 1869–1887, doi: 10.2458/azu_js_rc.55.16947, 2013.

[REI 20] REIMER P.J., BAILLIE M.G.L., BARD E. et al., "The IntCal20 Northern Hemisphere radiocarbon age calibration curve (0-55 cal KBP)", *Radiocarbon*, pp. 1–33, doi: 10.1017/RDC.2020.41, 2020.

[REI 21] REICHE I., BECK L., CAFFY I., "New results with regard to the Flora bust controversy: Radiocarbon dating suggests nineteenth century origin", *Scientific Reports*, vol. 11, p. 8249, doi: 10.1038/s41598-021-85505-x, 2021.

[REN 15] RENAUD P., Les essais atmosphériques d'armes nucléaires : des retombées radioactives à l'échelle planétaire, available at: https://www.irsn.fr/sites/default/files/retombees-tirs-armes-nucleaires/pdf, 2015.

[RIC 13a] RICHARDIN P., GANDOLFO N., "Datation et authentification des œuvres de musée – apport de la datation par le carbone 14", *Spectra Analyse*, vol. 292, pp. 55–60, 2013.

[RIC 13b] RICHARDIN P., GANDOLFO N., "Radiocarbon dating and authentication of objects from ethnographic museums", *Radiocarbon*, vol. 55, nos 3–4, pp. 1810–1818, 2013.

[RIC 17] RICHARDIN P., PERRAUD A., HERTZOGET J. et al., "Radiocarbon dating of a series of the heads of Egyptian mummies from the Musée des Confluences, Lyon (France)", *Radiocarbon*, vol. 59, pp. 609–619, 2017.

[RIC 21] RICHARDIN P., "La datation par le carbone 14 s'invite dans nos musées", *Technè*, vol. 52, pp. 51–59, doi: 10.4000/techne.9869, 2021.

[ROG 82] ROGERS D.W.O., "More realistic Monte Carlo calculations of photon detector response functions", *Nuclear Instruments and Methods in Physics Research*, vol. 199, pp. 531–537, 1982.

[ROS 66] ROSHOLT J.N., DOE B.R., TATSUMOTO M., "Evolution of the isotopic composition of uranium and thorium in soil profiles", *Geological Society of America Bulletin*, vol. 77, pp. 987–1004, 1966.

[ROS 82] ROSHOLT J.N., "Mobilization and weathering", in IVANOVITCH M., HARMON R.S. (eds), *U-Series Disequilibrium: Applications to Environmental Problems*, Clarendon Press, Oxford, 1982.

[ROS 21] ROSIER C., Une naine brune, nommée "L'accident", intrigue les astronomes, available at: https://www.rtbf.be/detail_une-naine-brune?id=1083692, 2021.

[ROU 11] ROUVET J., Détermination de la pureté radiochimique de préparation radiopharmaceutiques par chromatographie liquide haute performance (HPLC) (mise en place du contrôle qualité du ^{99m}Tc-mertiatide (MAG3®) et étude de radiotraceurs innovants), PhD Thesis, Université de Rouen, Rouen, 2011.

[ROU 13] ROUSSEL J., L'effet Doppler, available at: https://femto-physique.fr/presentation-doppler.pdf, 2013.

[ROU 17] ROUDIL J., Influence de la pathologie de Lewy sur le phénotype clinique de la maladie d'Alzheimer, PhD Thesis, Université de Lorraine, Lorraine, 2017.

[SAB 19] SABA D., Les radionucléides et le plomb émis par une usine de fertilisants au Liban : étude de leurs mobilités dans les sols et les plantes pour une évaluation du risque sanitaire, PhD Thesis, Université Paris-Saclay, Paris, 2019.

[SAC 58] SACKET W.M., Ionium–uranium ratios in marine deposited calcium carbonates and related materials, PhD Thesis, Washington University, St Louis, 1958.

[SAC 60] SACKETT W.M., "The protactinium-23 1 content of ocean water and sediments", *Science*, vol. 132, pp. 1761–1762, 1960.

[SAK 96] SAKHO I., DIOP A., GAYE B. et al., Chromatographie ionique avec détection conductimétrique. Application à la détermination quantitative des anions inorganiques dans les eaux de distribution urbaine du Sénégal, 2e journées annuelles de la société ouest africaine de chimie, Dakar, 1996.

[SAK 97] SAKHO I., Détermination des anions inorganiques dans les eaux de consommation de trois régions du Sénégal (Dakar, Thiès et Diourbel) par la Chromatographie ionique, Diploma of Advanced Studies in Physical Chemistry Applied to Energy, Université Cheikh Anta Diop de Dakar, Dakar, 1997.

[SAK 11] SAKHO I., *Histoire de l'atome, de l'insécabilité au modèle probabiliste*, Éditions Publibook, Paris, 2011.

[SAK 16a] SAKHO I., *Guides Pratiques du Lycéen, Physique Terminales S*, Nouvelles Éditions Africaines du Sénégal, Dakar, 2016.

[SAK 16b] SAKHO I., "Energy dissipated by an aster accelerated in a gravitational field: Estimation of the lifetime of a planet or a star being destroyed", *Journal of Astrophysics and Aerospace Technology*, vol. 4, pp. 1–5, 2016.

[SAK 17] SAKHO I., Interaction rayonnement-matière, Master's Thesis in Materials Physics, Université Assane Seck de Ziguinchor, Senegal, 2017.

[SAK 19] SAKHO I., *Introduction to Quantum Mechanics 1: Thermal Radiation and Experimental Facts Regarding the Quantization of Matter*, ISTE Ltd, London and John Wiley & Sons, New York, 2019.

[SAK 20] SAKHO I., *Physique atomique, Systèmes hydrogénoïdes et systèmes héliumoïdes, Cours & exercices corrigés*, Éditions Ellipses, Paris, 2020.

[SAK 22] SAKHO I., *Nuclear Physics 1: Nuclear Deexcitations, Spontaneous Nuclear Reactions*, ISTE Ltd, London and John Wiley & Sons, New York, 2022.

[SAK 23a] SAKHO I., "Origine des éléments chimiques : du Big Bang à la nucléosynthèse explosive", *Conférence grand public. Journées scientifiques de la Commission de recherche de l'UFR SAT/UGB*, Université Gaston Berger de Saint Louis, Saint Louis, 15 March 2023.

[SAK 23b] SAKHO I., "Sakholian radius-to-mass ratio postulate applied to the calculation of the mass or the radius of a satellite in the Solar System and in the Milky Way", *International Astronomy and Astrophysics Research Journal*, vol. 5, no. 1, pp. 189–200.

[SAL 20] SALAUN P.-Y., Instrumentation en médecine nucléaire, UBO, Brest, available at: https://cerf.radiologie.fr/.

[SAN 07] SANTOS F.O., Diffusion élastique résonante, Diffusion inélastique et réactions astrophysiques, GANIL (Grand Accélérateur National d'Ions Lourds), available at: https://www.osti.gov/etdeweb/, 2007.

[SAN 12] SANCHEZ-CABEZA, J.A., RUIZ-FERNÁNDEZ, A.C., "210Pb sediment radiochronology: An integrated formulation and classification of dating models", *Geochimica et Cosmochimica Acta*, vol. 82, pp. 183–200, 2012.

[SAS 15] SASCO R., Développement d'un outil chronostratigraphique pour les archives climatiques : datations absolues (K/Ar,Ar/3Ar) et paléomagnétisme appliqués aux laves, PhD Thesis, Université Paris Sud-Paris XI, Paris, 2015.

[SAZ 13] SAZY M.A.B., Mesure des anisotropies de polarisation du fond diffus cosmologique avec l'interféromètre bolométrique QUBIC, PhD Thesis, Université Paris-Diderot – Paris VII, Paris, 2013.

[SCA 11] SCACCHI G., Nucléosynthèse. Formation des éléments chimiques dans l'Univers. Vie et mort des étoiles, ALS 8/12/2011, available at: http://als.univ-lorraine.fr/files/conferences/2011/nucleosynthese.pdf, 2011.

[SCH 07] SCHNEIDER J.A., ARVANITAKIS Z., BANG W. et al., "Mixed brain pathologies account for most dementia cases in community-dwelling older persons", *Neurology*, vol. 69, pp. 2197–2204, doi: 10.1212/01.wnl.0000271090.28148.24, 2007.

[SCH 08] SCHOLZ D., HOFFMANN D., "230Th/U-dating of fossil corals and speleothems", *Quaternary Science Journal*, vol. 57, nos 1–2, pp. 52–76, 2008.

[SCH 21] SCHER N., L'espérance de vie du cancer de la prostate de stade avancé, available at: https://radiotherapie-hartmann.fr/actualites/cancer-prostate/, 2021.

[SÉG 15] SÉGALA G., L'angiogenèse du tissu tumoral, available at: https://www.futura-sciences.com/sante/, 2015.

[SÈN 19a] SÈNE M., NDEYE M., "Quantifying fossil fuel CO_2 component over Dakar from 1960 to 2010 by radiocarbon observation in atmospheric CO_2 and using Mauna Loa as background", *International Journal of Energy and Environmental Science*, vol. 3, no. 3, pp. 52–60, 2019.

[SÈN 19b] SÈNE M., NDEYE M., "La méthode du carbone 13 dans le monitoring de l'émission de dioxyde de carbone fossile dans l'atmosphère", *Bulletin IFAN, Sér. A*, vol. 53, no. 1, pp. 117–136, Dakar, 2019.

[SÈN 19c] SÈNE M., Le carbone 13 dans l'environnement. Contribution à l'étude de la pollution atmosphérique par le dioxyde de carbone dans les zones urbaines. Cas de la région de Dakar, PhD Thesis, Université Cheikh Anta Diop de Dakar, Dakar, 2019.

[SEN 21] SENEWEB, Alzheimer : 1,5 million de sénégalais dans l'oubli, available at: https://www.seneweb.com/news/Societe/alzheimer-1-5-million-de-senegalais-dans-l-oubli_n_52565.html, 2021.

[SEP 08] SEPULVEDA A., SCHULLER P., WALLING D.E. et al., "Use of ^{7}Be to document soil erosion associated with a short period of extreme rainfall", *Journal of Environmental Radioactivity*, vol. 99, pp. 35–39, doi: 10.1016/j.jenvrad.2007.06.010, 2008.

[SHA 05] SHARP W.D., RENNE P.R., "The 40Ar/39Ar dating of core recovered by the Hawaii Scientific Drilling Project (phase 2), Hilo, Hawaii", *Geochemistry Geophysics Geosystems*, vol. 6, p. Q04G17, doi: 10.1029/2004GC000846, 2005.

[SHE 85] SHEN G.J., Datation des planchers stalagmitiques de sites acheuléens en Europe par les méthodes des déséquilibres des familles de l'uranium et contribution méthodologique, PhD Thesis, Museum National d'Histoire Naturelle et Université Pierre et Marie Curie, Paris VI, Paris, 1985.

[SHE 95] SHEN G.T., DUNBAR R.B., "Environmental controls on uranium in reef corals", *Geochimica et Cosmochimica Acta*, vol. 59, no. 10, pp. 2009–2024, 1995.

[SLI 13] SLIPHER V.M., "The radial velocity of the Andromeda Nebula", *Lowell Observatory Bulletin*, vol. 58, no. 2, 1913.

[SOC 21] SOCIÉTÉ FRANÇAISE D'ÉNERGIE NUCLÉAIRE, La médecine nucléaire, available at: https://www.sfen.org/energie-nucleaire/energie-service-progres/medecine-nucleaire, 2021.

[STE 77] STEIGER R.H., JÄGER E., "Subcommission on geochronology: Convention on the use of decay constants in geo- and cosmochronology", *Earth and Planetary Science Letters*, vol. 5, pp. 320–323, 1977.

[STI 09] STIRLING C.H., ANDERSEN M.B., "Uranium-series dating of fossil coral reefs: Extending the sea-level record beyond the last glacial cycle", *Earth and Planetary Science Letters*, vol. 284, nos 3–4, pp. 269–283, 2009.

[STU 98] STUIVER M., REIMER P.J., "CALIB Rev 3.3 (Data set 2)", *Radiocarbon*, vol. 30, pp. 1031–1083, doi: 10.1017/S0033822200019123, 1998.

[SWA 82] SWART P.K., HUBBARD J., "Uranium in scleractinian coral skeletons", *Coral Reefs*, vol. 1, no. 1, pp. 13–19, 1982.

[SYN 07] SYNAL H.A., STOCKER M., SUTER M., "MICADAS: A new compact radiocarbon AMS system", *Nuclear Instruments and Methods in Physics Research Section B*, vol. 259, pp. 7–13, 2007.

[TAL 03] TALBOT J.N., Guide pour la rédaction de protocoles pour la tomographie par émission de positons (TEP) au [18F]-fludésoxyglucose ([18F]-FDG) en cancérologie, Société Française de Biophysique et de Médecine Nucléaire (SFBMN), Paris, Version 1.0, available at: www.sfbmn.org, 2003.

[TAT 15] TATISCHEFF V., "La nucléosynthèse stellaire", *25e Festival d'Astronomie de Fleurance*, available at: http://fermedesetoiles.com/documents/supports/la-nucleosynthese-stellaire.pdf, 2015.

[THI 20] Thiebaux A., Glycémie (à jeun) : taux normal dans le sang, élevée, basse, available at: https://sante.journaldesfemmes.fr/fiches-anatomie-et-examens/2423998-glycemie-a-jeun-taux-normal-definition-elevee-basse-comment-baisser/, 2020.

[THO 18] Thomas M.J., Port du tablier plombe : étude dosimétrique, PhD Thesis. Université Lyon 1, Lyon, 2018.

[THU 62] Thurber D.L., "Anomalous ^{233}U/^{238}U in nature", *Geophysical Research Letters*, vol. 67, pp. 3518–3520, 1962.

[THU 63] Thurber D.L., Natural variation in the ratio ^{233}U/^{238}U and investigation of the potential of ^{233}U for pleistocene chronology, PhD Thesis, University of Columbia, 1963.

[TRE 18] Treuil J.-P., L'idée du Big-Bang : histoire de la construction progressive du modèle standard de la cosmologie, available at: http://scienceinter.com/construction%20modele%20standard.pdf, 2018.

[TRE 22] Tremblay L., 10 – Spectres et diagramme HR, Collège Mérici, Quebec, Version 2022, available at: https://physique.merici.ca/astro/chap10astsol.pdf, 2022.

[TUR 66] Turner G., Miller J.A., Grasty R.L., "The thermal history of the Bruderheim meteorite", *Earth and Planetary Science Letters*, vol. 1, pp. 155–157, 1966.

[TYN 69] Tyndall J., *Le son : cours expérimental fait à l'Institution Royale*, Gauthier-Villars, Paris, available at: https://www.abebooks.com, 1869.

[UNI 18] Union Nationale des Associations France Alzheimer et maladies apparentées, La maladie à corps de Lewy, available at: https://www.francealzheimer.org/2018/01/Brochure-Maladie-%C3%A0-corps-de-Lewy, 2018.

[VAC 18] Vachez Y., Troubles neuropsychiatriques de la maladie de Parkinson et stimulation haute fréquence du noyau subthalamique : approche préclinique chez le rat de l'hypothèse dopaminergique de l'apathie, PhD Thesis, Université Grenoble Alpes, Grenoble, 2018.

[VAL 05] Valladas H., Tisnérat-Laborde N., Cacher H. et al., "Bilan des datations carbone 14 effectuées sur des charbons de bois de la grotte Chauvet", *Bulletin de la société préhistorique française*, vol. 102, no. 1, pp. 109–113, 2005.

[VAL 09] Vala C., Synthèse de groupements prosthétiques glucidiques : vers de nouveaux traceurs peptidiques pour l'imagerie par Tomographie par émission de positons (TEP), PhD Thesis, Université Henri Poincaré, Nancy-Université, Nancy, 2009.

[VAN 14] Van Strydonck M., Bénazeth D., "Four coptic textiles from the Louvre collection ^{14}C redated after 55 years", *Radiocarbon*, vol. 56, no. 1, pp. 1–5, doi: 10.2458/56.16787, 2014.

[VET 23] Vetopsy, Modèle standard des particules, comment les interactions sont portées par des particules ?, available at: http://www.vetopsy.fr/modele-standard-particules/interactions-fondamentales-particules.php, 2023.

[WAS 98] WASTIEL C., KOSINSKI M., "Contrôle de qualité des produits radiopharmaceutiques par chromatographie liquide HPLC", *Analusis Magazine*, vol. 26, no. 2, pp. M18–M22, available at: https://analusis.edpsciences.org/articles/analusis/pdf/1998/02/m080298.pdf, 1998.

[WHO 22] WORLD HEALTH ORGANIZATION, Parkinson disease, available at: https://www.who.int/news-room/fact-sheets/detail/parkinson-disease, 2022.

[WHO 23] WORLD HEALTH ORGANIZATION, Dementia, available at: https://www.who.int/news-room/fact-sheets/detail/dementia, 2023.

[WIT 03] WITKOWSKI N, *Une histoire sentimentale des sciences*, Le Seuil, 2003. https://sirius.nathan.fr/9782091723761/assets/chapitre-3-exercice-resolu/preview.

[ZIM 06] ZIMMERMANN R., "La médecine nucléaire. La radioactivité au service du diagnostic et de la thérapie", *EDP Sciences*, available at: http://livre21.com/LIVREF/F5/F005136.pdf, 2006.

Index

Other titles from

in

Waves

2021

DAHOO Pierre-Richard, LAKHLIFI Azzedine
Infrared Spectroscopy of Symmetric and Spherical Top Molecules for Space Observation 1
(Infrared Spectroscopy Set – Volume 3)
Infrared Spectroscopy of Symmetric and Spherical Top Molecules for Space Observation 2
(Infrared Spectroscopy Set – Volume 4)

2020

DANIELE Vito G., LOMBARDI Guido
Scattering and Diffraction by Wedges 1: The Wiener-Hopf Solution - Advanced Applications
(Waves and Scattering Set – Volume 1)
Scattering and Diffraction by Wedges 2: The Wiener-Hopf Solution - Advanced Applications
(Waves and Scattering Set – Volume 2)

SAKHO Ibrahima
Introduction to Quantum Mechanics 2: Wave-Corpuscle, Quantization & Schrödinger's Equation

2019

BERTRAND Pierre, DEL SARTO Daniele, GHIZZO Alain
The Vlasov Equation 1: History and General Properties

DAHOO Pierre-Richard, LAKHLIFI Azzedine
Infrared Spectroscopy of Triatomics for Space Observation
(Infrared Spectroscopy Set – Volume 2)

RÉVEILLAC Jean-Michel
Electronic Music Machines: The New Musical Instruments

ROMERO-GARCIA Vicente, HLADKY-HENNION Anne-Christine
Fundamentals and Applications of Acoustic Metamaterials: From Seismic to Radio Frequency
(Metamaterials Applied to Waves Set – Volume 1)

SAKHO Ibrahima
Introduction to Quantum Mechanics 1: Thermal Radiation and Experimental Facts Regarding the Quantization of Matter

2018

SAKHO Ibrahima
Screening Constant by Unit Nuclear Charge Method: Description and Application to the Photoionization of Atomic Systems

2017

DAHOO Pierre-Richard, LAKHLIFI Azzedine
Infrared Spectroscopy of Diatomics for Space Observation
(Infrared Spectroscopy Set – Volume 1)

PARET Dominique, HUON Jean-Paul
Secure Connected Objects

PARET Dominque, SIBONY Serge
Musical Techniques: Frequencies and Harmony

RÉVEILLAC Jean-Michel
Analog and Digital Sound Processing

STAEBLER Patrick
Human Exposure to Electromagnetic Fields

2016

ANSELMET Fabien, MATTEI Pierre-Olivier
Acoustics, Aeroacoustics and Vibrations

BAUDRAND Henri, TITAOUINE Mohammed, RAVEU Nathalie
The Wave Concept in Electromagnetism and Circuits: Theory and Applications

PARET Dominique
Antennas Designs for NFC Devices

PARET Dominique
Design Constraints for NFC Devices

WIART Joe
Radio-Frequency Human Exposure Assessment

2015

PICART Pascal
New Techniques in Digital Holography

2014

APPRIOU Alain
Uncertainty Theories and Multisensor Data Fusion

JARRY Pierre, BENEAT Jacques N.
RF and Microwave Electromagnetism

LAHEURTE Jean-Marc
UHF RFID Technologies for Identification and Traceability

SAVAUX Vincent, LOUËT Yves
MMSE-based Algorithm for Joint Signal Detection, Channel and Noise Variance Estimation for OFDM Systems

THOMAS Jean-Hugh, YAAKOUBI Nourdin
New Sensors and Processing Chain

TING Michael
Molecular Imaging in Nano MRI

VALIÈRE Jean-Christophe
Acoustic Particle Velocity Measurements using Laser: Principles, Signal Processing and Applications

VANBÉSIEN Olivier, CENTENO Emmanuel
Dispersion Engineering for Integrated Nanophotonics

2013

BENMAMMAR Badr, AMRAOUI Asma
Radio Resource Allocation and Dynamic Spectrum Access

BOURLIER Christophe, PINEL Nicolas, KUBICKÉ Gildas
Method of Moments for 2D Scattering Problems: Basic Concepts and Applications

GOURE Jean-Pierre
Optics in Instruments: Applications in Biology and Medicine

LAZAROV Andon, KOSTADINOV Todor Pavlov
Bistatic SAR/GISAR/FISAR Theory Algorithms and Program Implementation

LHEURETTE Eric
Metamaterials and Wave Control

PINEL Nicolas, BOURLIER Christophe
Electromagnetic Wave Scattering from Random Rough Surfaces: Asymptotic Models

SHINOHARA Naoki
Wireless Power Transfer via Radiowaves

TERRE Michel, PISCHELLA Mylène, VIVIER Emmanuelle
Wireless Telecommunication Systems

2012

LALAUZE René
Chemical Sensors and Biosensors

LE MENN Marc
Instrumentation and Metrology in Oceanography

LI Jun-chang, PICART Pascal
Digital Holography

2011

BECHERRAWY Tamer
Mechanical and Electromagnetic Vibrations and Waves

BESNIER Philippe, DÉMOULIN Bernard
Electromagnetic Reverberation Chambers

GOURE Jean-Pierre
Optics in Instruments

GROUS Ammar
Applied Metrology for Manufacturing Engineering

LE CHEVALIER François, LESSELIER Dominique, STARAJ Robert
Non-standard Antennas

2010

BEGAUD Xavier
Ultra Wide Band Antennas

MARAGE Jean-Paul, MORI Yvon
Sonar and Underwater Acoustics

2009

BOUDRIOUA Azzedine
Photonic Waveguides

BRUNEAU Michel, POTEL Catherine
Materials and Acoustics Handbook

DE FORNEL Frédérique, FAVENNEC Pierre-Noël
Measurements using Optic and RF Waves

FRENCH COLLEGE OF METROLOGY
Transverse Disciplines in Metrology

2008

FILIPPI Paul J.T.
Vibrations and Acoustic Radiation of Thin Structures

LALAUZE René
Physical Chemistry of Solid-Gas Interfaces

2007

KUNDU Tribikram
Advanced Ultrasonic Methods for Material and Structure Inspection

PLACKO Dominique
Fundamentals of Instrumentation and Measurement

RIPKA Pavel, TIPEK Alois
Modern Sensors Handbook

2006

BALAGEAS Daniel *et al.*
Structural Health Monitoring

BOUCHET Olivier *et al.*
Free-Space Optics

BRUNEAU Michel, SCELO Thomas
Fundamentals of Acoustics

FRENCH COLLEGE OF METROLOGY
Metrology in Industry

GUILLAUME Philippe
Music and Acoustics

GUYADER Jean-Louis
Vibration in Continuous Media

Printed and bound by CPI Group (UK) Ltd, Croydon, CR0 4YY

11/08/2024

14538831-0001